BY **ANY** MEANS NECESSARY!

An Entrepreneur's Journey into Space

D1622131

By **GREGORY H. OLSEN**

with THOMAS V. LENTO

GHO Ventures, LLC • Princeton, NJ

©2009 by Gregory Olsen

International Standard Book Number (ISBN): 978-0-615-31101-2
Library of Congress Control Number: 2009939681

Contents

This book is dedicated to the most special people in my life:
daughters Kimberly and Krista plus grandchildren
Justin, Carter, Romina, Danielle, Athina and Oriana.
They make the sun shine for me.

Foreword

I can summarize how I feel about my life in four words: "*Average guy makes good.*" Now I know I'm not what most people would call average, considering that I've managed to parlay success in a technology business into a stay on the International Space Station.

But in truth I don't believe that this achievement was the result of any special kind of genius. In fact I don't do anything extremely well. There are a very few things I do very well, like working hard, and a few more things I can do moderately well.

But I'm only marginally adequate at the vast majority of life's activities. And there are many areas where I'm below average. These would include my golf game, my scholastic record, how I've managed companies, even my relationships.

Sometimes I wonder how it all happened. When I was 17 and struggling mightily with my first calculus course at college, I felt really inadequate as I watched foreign students performing difficult "integrals" like they were adding 2+2. I still get that feeling at age 64 when I meet venture capitalists who run on about "business models" and "value propositions." When I meet other successful entrepreneurs, I often feel that they have skills beyond my own. And you can imagine how I felt among experienced astronauts and cosmonauts as I trained for my space mission: a complete rookie who wasn't even an amateur pilot!

I guess one of the attributes that have gotten me to where I am is tenaciousness. Even when I feel inadequate to the task before me, I somehow manage to push on with it. And in most cases, somehow manage to pull it off.

So in another sense this book is dedicated to those who wonder what it takes to "succeed" or to "get ahead" or whatever they define as their goal in life. The secret is…there ain't no secret! You just have to put your head down and do it, in spite of the doubts, the setbacks, the failures, or how long it takes.

Fittingly enough, my favorite mantras are DON'T GIVE UP and JUST DO IT! I sincerely believe those words will get you over the rough spots. Whenever things are not going particularly well in my life, which can happen anytime, anyplace, to anyone, I reflect back on them. It may not make me feel any better – usually it doesn't – but it gives me a reason to go on in the hope that those good times and good feelings of the past will eventually come back.

This is not a "tell-all" book. I'm no saint, and like most people I've done things in my life that I'd just as soon forget about. But by today's tabloid standards, they would hardly make the back page, so there's not much point in retelling them. Plus, I'm a very private person and don't particularly care to share the details of my life with many people.

The purpose of this book is not to write a history of Greg Olsen, even though it has come out sounding that way, now that I reread it. Instead, my goal is to share the experiences that I have had in life and hope that reading about them might encourage other people to say, "Hey – if this guy can do it…maybe I can too!"

My thanks to novelist, poet, and producer Sandra Hochman for suggesting the title of this book.

By Any Means Necessary!

An Entrepreneur's Journey into Space

By Gregory H. Olsen

with Thomas V. Lento

Soyuz rocket blasts off.

1
On the Launchpad

N.J. Scientist prepares to shoot for the stars as space tourist
Baikonur, Kazakhstan (Oct. 1, 2005) — The U.S. scientist slated to become the third non-astronaut to visit the international space station said Friday he was nervous ahead of the launch and defended his joining the expedition as a necessary step in the evolution of space flight.

"I will feel most relaxed and most happy when the rocket is taking off," said Gregory Olsen, whose trip reportedly costs $20 million. "This has been two years of very hard work. In 20 hours, I will feel very, very good."

Blastoff is scheduled for today at 7:54 a.m. Moscow Time.

— Mike Eckel, The Associated Press

By launch time the nerves were gone. They had never been about the flight anyway – they were rooted in the nagging anxiety that something might happen to keep me from going into space. The medical staff at Star City, the Russian space center, had blacklisted me from the program twice, and they could do it again.

But it was the morning of liftoff at last. I was strapped into Soyuz TMA-7, the launch vehicle, along with NASA astronaut Bill McArthur and our commander, Russian cosmonaut Valeri Tokarev, awaiting the countdown.

In just a few moments we would be leaving on a 10-day trip that would be literally out of this world. We were going to the International Space Station, floating in orbit 240 miles above the surface of the earth.

As a crew we knew what to expect. The Soyuz would take us to earth orbit in just eight or nine minutes from initial ignition.

For two minutes the rocket's first two stages put out nearly 500,000 pounds of thrust, lifting the ship off the ground and starting it on its way.

1

In the next six minutes the second and third stages sling the vehicle out of earth's atmosphere toward low-earth orbit.

The third stage, including the crew capsule, manages the last part of the trip, to the International Space Station and home again.

We had been in the capsule for over two hours and my bladder was starting to feel the effects of two cups of tea. We were wearing "Huggie" disposable diapers for just this reason, but I feared it might smell up the cabin. I leaned over to Tokarev and in broken Russian blurted out, "Valeri … ya toilet." In much better English he replied, "Don't worry, Greg. I already went." It made me feel so much better as I relieved myself.

But that was nothing next to the elation I was also experiencing. At last, after more than two years of hard work, rejection, and setbacks, my dream was coming true. I felt calm, serene, at peace with myself, ready to blast off!

The fact that my daughter Krista and her 4-year-old son Justin, plus 30 other friends and family, were there to watch the launch only sweetened the pleasure.

How did I get there? What did it mean? It was certainly a long way from Brooklyn, NY where I was born, and New Jersey, where I grew up and have spent most of my life. Certainly no one who knew me as a kid in Ridgefield Park would have believed that Greg Olsen was going to be successful enough one day to book a flight on a spaceship.

In fact there were a lot of things about my life that would have surprised them. Earning a doctorate in a scientific field. Starting my career in South Africa. Doing scientific research. Founding two successful companies.

It would all seem so unlikely to the people who knew me way back when. As I sat waiting for the countdown, a smile came over my face as I flashed back to my younger days.

Flunking trig

I'd been a ne'er-do-well kid, often in trouble with teachers or the police, with no ambitions other than to smoke, get a car, and have a girlfriend. My high school principal had suspended me twice, snarling that I would never get into ANY college with the record I had. He guaranteed it. And he was almost right.

I hung around with two other guys, Roger Belisle and Bobby Comiso. We were known as the "COB" – Comiso, Olsen and Belisle, dedicated to annoying people ("bugging," as we called it) and generally doing outrageous things, pulling pranks – some not so harmless (like throwing hard snowballs at a teacher's head). Later, Danny Boswell rounded us out to "COBB". We had no other goals or ambitions, other than to make life difficult for people we decided we didn't like.

When graduation came, I was sitting with a 78 average and had just failed trigonometry for the entire year. Ol' Mrs. Petix – she had her method of solving problems and if you didn't like it, well, too bad! Well, I didn't like it, and furthermore I wasn't interested. Not doing the homework seemed sensible to me at the time. I was going to show her…only, she wound up showing me.

Facing what seemed like limited prospects for a rosy future, right after graduation a couple of high school buddies and I went to take the entrance exam for the US Army.

This was in 1962, before the big plunge into Vietnam. The Army was seen as a "fallback" choice if you had no idea what to do after high school. Buddies would come back from service saying how easy Army life was. You would get posted to Germany or Korea… have some good times…see the world.

"You did very well, son," the recruiter told me when it was over. "I can arrange for you to go almost anywhere you like. Just get your parents to sign this permission slip." (Since I had only just turned 17 when I graduated from high school, I needed parental approval to enlist in the service.)

So I raced home to get the required signature and begin the next stage of my life. After the Army, I assumed I'd go into my Dad's union as an apprentice electrician. I had my life all figured out.

And none of it turned out the way I expected.

I thought I could plan my future the way engineers and scientists plot the course of a rocket launch. First you lift off, leaving earth the way I graduated from high school (well, hopefully a little better than that). The first-stage booster burns out and drops off at a specific time, and you leave it behind, the way I'd leave the Army.

Then the second stage, in my case becoming an apprentice electrician, takes you to your initial orbit. After that it's an easy journey from one part of life to another.

But as I discovered after the Army exam, and many more times since then, life is unpredictable. It has a way of changing directions on you when you least expect it.

I showed the permission slip to my Dad. He was a construction electrician with Local 3 IBEW (International Brotherhood of Electrical Workers, the electricians' union) in New York City. He had hands like vicegrips with a personality to match. He was old school, hard-assed, my way or the highway. You didn't "debate" or reason with him.

How tough was he? When I was around five years old and we were living in Brooklyn, he and his friends would go to O'Sullivan's tavern in Bay Ridge every day after work for some socializing. I thought it was perfectly normal for grown men to have three Rheingold beers and two shots of Seagram's 7 before dinner every night. (I revisited O'Sullivan's just recently and it's still there on 3rd Ave and 89th Street. Frank O'Sullivan, son of the owner and about my age, still remembers those old days!)

My Mom would send me down there to drag him home for dinner. He usually wasn't ready to leave when I got to the tavern. While I hung around waiting for him to finish up, his construction buddies would give me nickels to play the juke box. They'd also let me keep the pennies in change packed under the cellophane wrapping of cigarettes they bought from the vending machine. After one or two frantic phone calls from my mother ("...tell her I a'ready left...") we would walk home together.

With Frank O'Sullivan at O'Sullivan's Tavern in Bay Ridge, Brooklyn, where my Dad used to hang out.

Once he carried me home on his shoulders. There aren't too many other happy moments with him that I can recall. So in 1962, at age 17, I didn't expect him to willingly sign my permission slip. A more probable outcome would be for him to say, "Why do you want to do something as stupid as this?" An argument would ensue, and I'd storm off to forge his signature on the paper.

But Dad, like life, proved full of surprises. When I approached him about enlistment, he slowly dropped his reading glasses, and said softly, "You know, it's getting harder to get into the union now. There's talk about requiring six months of college to become an apprentice. So why don't you try college for one semester? If you don't like it, I'll sign the paper."

It was SO reasonable, so unlike him, that I heard myself mumble "Well....OK."

This was my first real introduction to a basic reality of human existence. In the words of John Lennon, "Life is what happens to you while you're busy making other plans."

Like anyone who's facing a major turning point in life, I was simply trying to plan my direction as best as I could. Maybe my plan wasn't all that well thought out, but at least I felt like I had a sense of where I was going and how to get there.

Fortunately I discovered that it was good to be flexible, too. If somebody suggests a better idea or you stumble across a new opportunity, you have to be ready to rethink your position. When my Dad laid out an alternate path at the very last minute, one that offered more options than just joining the army, I was smart enough – or perhaps more accurately, surprised enough – to take it.

That's only the first of many times I've been heading down one road when a surprising turn of events changed my direction and radically reshaped my future.

I never did join the army. While there were lots of stops and detours along the way, Dad's response set me on a path that eventually took me through college and into the world of technology and business. Ultimately it brought me to this seat on a rocket some 43 years later.

My Dad didn't live to see me reach for the stars – too many unfiltered Camel cigarettes got him in the end. So in gratitude for the start he gave me, I had his old set of keys from Local 3 with me on the Soyuz. It was a way of letting him share the adventure.

Flight prep

It took two years of hard physical training, Russian language lessons, and technical study to get ready for the trip to the space station. The work was demanding and there were some glitches along the way – a health problem nearly washed me out of the program for good – but it had all been worth it.

Here I was, waiting for liftoff. It almost seemed like my whole life had pointed toward this moment.

Don't get me wrong. It's not like I always knew that someday I'd be sitting on top of a rocket out on the steppes of Kazakhstan, waiting for the countdown to go from **десять** (ten, pronounced "dyes-et") to "**ноль**" (you guessed it, zero, pronounced "nul." Nobody can see that far ahead.

By the way, that countdown was me ticking off the seconds in my head. The Russians don't do an actual countdown the way NASA does – they just go!

But the idea of space flight had fascinated me since the seventh grade. And even though I didn't design rocket ships for a living, a career in science, engineering, and business had given me a real appreciation for the technical challenges of putting people into space.

It also gave me the financial resources to book a seat in this great

Up in space with my Dad's keys from Local 3.

human enterprise. The privilege cost a lot – in the $20 million range. But I'd given away more than that to charity over the years, so I didn't think I was neglecting my responsibility to make the world a better place.

Besides, this wasn't a passive tourist trip just for personal pleasure. I was going to do experiments in space to help advance science. Plus the trip opened to door for me to give over 300 space talks to school kids, trying to get more of them into science and engineering careers.

Was it luck that got me here? The truth is, luck does play a big role in life. And believe me, I've had more than my share. I'd had the great good fortune of being steered toward college, where I found my true calling in science. Circumstance had put me in a position to start and run two successful companies, and to profit handsomely from their sale.

I'm not very religious, but there are thoughts and concepts in many religions that are truly valid and universal. One of these is the Lutheran concept of God's grace. This is the belief that no one really deserves or can earn God's love or forgiveness. We just get these things through his grace, as free gifts to us.

Now, I'm not going to debate the fine points of Lutheran theology, but I will tell you this: "grace" in the sense of a gift, good fortune, some lucky break you haven't done much to "earn" – this has pervaded my life. Something always seems to come along that leads to success or happiness.

I guess that's what makes me such an optimist. I have this inborn belief that somehow, someway, things will work out. I don't know how, and I usually suffer the anxiety and depression that everyone else feels when things don't seem to be going well. But I just put my head down and keep on working until I find a way to improve the situation.

Often, this involves barely getting by, like passing a course you're not doing well in, getting a letter in football after marginal play on special teams, or just "scraping by" on my PhD general exams.

Work, of course, is the other part of the equation. If you just go ahead and get the job done in spite of your doubts, things come out right more often than not. Hard work pays off in the end. As a certified workaholic I find it easy to get so absorbed in a task that I can shut out distractions until I find a solution, or it finds me. Grace and work: an unbeatable combination.

Don't get me wrong. I know I've been lucky in life – very lucky, in fact. But I've come to realize that luck is often better described as "opportunity," and it amazes me how many people get presented with opportunities and just let them slide by.

During the fiber optics boom of 2000 a fellow entrepreneur told me he'd just been offered $500 million for his company. "But I know we're worth more than that," he said. By the time this guy woke up the stock market had crashed and he wound up with nothing.

There were plenty of other examples as well. These were people who had an opportunity, and never took advantage of it. If you hang in there and work hard, sooner or later opportunities, big or small, will present themselves. Your "luck" often depends on how well you respond.

Another variant of opportunity is money. (It seems to me that's what money is, an opportunity, not a solution.) Money can create more problems than it solves if it's treated as an end in itself. But more on this later....

On the Launchpad

Soyuz TMA-7 stood in launch position. Outside the rocket the ground crew was finishing up the last procedures before final countdown. Inside Soyuz the three of us were cocooned in our space suits and strapped in our seats for blastoff.

Cocooned is a good word for it, by the way. The seats were exactly matched to our bodies for maximum comfort and protection. They're made from body molds. I was put into a bathtub full of wet plaster, then pulled out of it with an overhead crane as it hardened. That created the mold for my seat.

As for the space suit, it isn't a loose, baggy, one-size-fits-all contraption either. It too is custom-molded to the wearer's body, and it fits you like a glove. That's the only way it can protect you from the rigors of space.

The Russian "sokol" space suit, which is not used for spacewalks, weighs about 22 pounds and has a rubber lining to hold air pressure in case the cabin depressurizes. It also holds in your body heat. That's why you always see astronauts and cosmonauts carrying boxes out to the launchpad. The "boxes" are actually ventilating fans!

One thing a space suit doesn't protect you from is acceleration. After over 900 hours of training at Star City we knew what to expect. As the rocket accelerates toward space you're pressed further and further into your seat by a force that's about three times the gravity you experience on earth. At its peak a 200-pound man will weigh over 600 pounds. Your arms feel like they have 10-pound weights attached to them. It's a lot of stress, but it only lasts a few minutes.

We couldn't watch the preparations for launch, or for that matter see our surroundings, because we were on our backs in a capsule pointing straight up, and our view was blocked by the fairing that protects the capsule until the rocket is well off the ground. We wouldn't have a view of any kind until we were 40 miles or so above the ground. At that point the fairing would fall off with an explosive boom and reveal the beautiful blue sphere of the earth beneath us.

But we knew what the scene looked like outside the Soyuz. The Baikonur Cosmodrome, the launch site, is in Kazakhstan, not Russia, in the middle of a beautiful but desolate landscape. One of two Russian launch centers in 2005, the site was built when Kazakhstan was part of the Soviet Union. Since the Soviet breakup Russia has continued to run it under an arrangement with the Kazakh government.

Kazakhstan is a remote, sparsely populated land with long, flat stretches of steppes – arid semi-desert land, great for recovering spent rocket sections and returned space capsules. Launching over water was not feasible because Russia does not have great access to the sea, especially for the all-important easterly takeoffs. (Since the earth rotates to the east, launching easterly gives you a starting velocity of almost 1,000 miles per hour at the equator, or about 700 mph at the 45° latitude of Baikonur.)

Baikonur was considered so important during the Soviet era that the Russians deliberately mismarked the launch site on official maps in an attempt to hide it from foes (as if our spy satellites couldn't pick it up...).

Six months later, when I traveled along with the Russian search and rescue team to the landing site to welcome my crewmates Valeri Tokarev and Bill McArthur back to earth on their return from the ISS, I got a first-hand appreciation for the isolation of the terrain. The highlight of that trip was to give Bill a whiff of fresh-ground Starbucks coffee right after he landed!

Liftoff

Finally we were ready for launch. I couldn't wait for my chance to see earth from space. Yet oddly enough the excitement I felt didn't cause any major elevation in my pulse. Here I was, the oldest crew member by six years, yet I had the lowest heart rate. I was 60 years old, with a pulse that matched my age. During launch it would actually drop into the 50s. That pervasive sense of calm euphoria must have helped to relax me.

Most astronauts and cosmonauts like to claim some sort of record for their flights into space: "the longest … the first … the most." Well, I think my crewmates and I can lay claim to being the oldest crew ever to fly the Soyuz: ages 60, 54, and 53!

As the final seconds ticked off and I felt the rocket begin to shudder, I wished my arms were free so that I could punch my fists into the air and say: "YES – I finally did it!"

Greg Olsen, Valeri Tokarev and Bill McArthur (l to r) on their walk to the Soyuz Rocket.

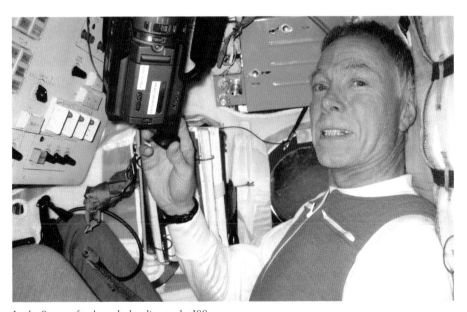

In the Soyuz after launch, heading to the ISS.

Swabbing to gather bacteria for my experiments for the European Space Agency.

2 A New Vision: Space Travel

Scientist Optimistic on Space Experiments
West Windsor, NJ (April 19, 2004) – When millionaire scientist Gregory Olsen turns space tourist months from now, he won't just be enjoying the view from the International Space Station…. Olsen plans to bring along infrared cameras made by his company…gathering data in space and possibly showing that the infrared detectors could be made more efficiently in gravity-free space – assuming commercial space travel eventually becomes affordable.

— Linda A. Johnson, The Associated Press

Breaking free of the bonds of gravity was an exotic, incredible feeling. There I was, rocketing toward the International Space Station.

My goal was not to be a "space tourist," but a working scientist with a job to do. It didn't make sense to me to turn a space trip into a ten-day vacation. After all I'd been through to get there, including over 900 hours of intense study and physical training, I wanted to be an active participant. Besides, it was too big an opportunity for a scientist to miss.

Originally I'd planned to do some experiments with cameras from Sensors Unlimited, the company I'd co-founded back in 1992. Talk about turning a space flight into a busman's holiday! But our government clients had concerns about security, and we weren't allowed to take the cameras with us.

I had also wanted to perform some crystal growth experiments using the NASA glovebox, which had a furnace. NASA had already grown an exotic material called indium antimonide, used to make heat-sensing cameras, during one of its flights.

My late crystal-growing buddy Bill Bonner and I wanted to add some gallium into the mix, which would make the cameras more like those made by Sensors Unlimited. Unfortunately the glovebox became unavailable, and the experiment never got off the ground.

So I agreed to conduct some experiments for the European Space Agency (ESA). My task was to study the response of the human body to microgravity. The ultimate purpose was to help uncover the causes of motion sickness and lower back pain. I also collected nearly 50 bacteria samples so that ESA could see what new strains might grow on the station.

There was something else I wanted to do, too: encourage young people to dream big and believe that anything is possible through hard work and dedication. I was also trying to motivate them to study science and technology.

What better way to do that than to let them share the wonder and awe of going into space? I could communicate the excitement of exploration, and show how fulfilling science can be, in addition to being a rewarding career.

I arranged for this part of my mission to start while I was still in space. We had a ham radio hookup at the Space Station that could make contact with Princeton University and Ridgefield High Schools in New Jersey, and Fort Hamilton High in Brooklyn, near my birthplace. It would let me chat with students about my experiences while they were happening.

They'd also hear how a career in science not only inspired me to go

On the HAM radio, speaking to students at Ridgefield Park High school.

into space, but also gave me the tools for success in business, which had made it possible to realize the dream.

All of that is true. But it's apparent only in hindsight. I never planned it that way.

In fact, many of the things I've done in life were the result of a sudden, unexpected encounter, not the outcome of a long-term plan. When the idea of going into space first entered my head, it was that kind of epiphany.

Space adventures

If you're ever looking for me in the morning in Princeton, check out the Starbucks coffee shop on Nassau Street. You can usually find me there with a big cup of coffee and the newspaper. I'm there so much I don't have to place an order – the staff knows what my "usual" is. I've met a lot of people and held a lot of meetings in Starbucks. I even funded a couple of companies there, including my first investment, Princeton Power Systems.

So it's not a surprise that I got the idea to go into space in that coffee shop. I know exactly when it happened: June 18, 2003. I still have the *New York Times* article that inspired me hanging on my wall. It described how a young company called Space Adventures Ltd. had

Reprinted From

The New York Times

Copyright © 2003 The New York Times NEW YORK, WEDNESDAY, JUNE 18, 2003 Page A20

For Those Who Can Afford It, 2 New Chances to Fly to Space

By JOHN
SCHWARTZ

An American company and the Russian space agency are to announce an agreement today to resume tourist flights to the International Space Station in 2004 or 2005, lifting a moratorium imposed after the Columbia shuttle disaster.

For $20 million each, travelers will have the chance to take part in what could be thought of as really extreme travel: a notoriously rough flight, cramped quarters and food that can make airline fare look appetizing. But the view is very, very nice.

Space Adventures, a company in Arlington, Va., devoted to space tourism, and the Russian Aviation and Space Agency said yesterday that they had reached an agreement on April 30 to

Space Agency will discuss with NASA and the other partners how this project can be conducted within the procedures that exist within the International Space Station partnership."

It is unclear whether both tourists will go up on a single flight and whether the flights will occur in 2004 or 2005. Sergei A. Gorbunov, a spokesman for the Russian space agency, said those details would be worked out later in the year.

Through a translator, Mr. Gorbunov said that the need to resupply the space station while the American shuttle fleet was grounded made it necessary to suspend space tourism.

The first tourist in space, Dennis A. Tito, an American businessman, was

It takes a special type of person to buy a trip into space — the kind of person who, for one thing, can find $20 million scattered between his sofa cushions. But Eric Anderson, the chief executive of Space Adventures, said a dozen potential customers had already gone far enough to begin qualifying medically.

Any traveler would have to be physically fit enough to pass the battery of required medical tests and would have to devote several months to a daunting training regimen.

Norman E. Thagard, a former NASA astronaut who was the first American to fly in a Soyuz rocket and who is now an adviser to Space Adventures, said that high-priced pioneering had always been arranged for two private citizens, Dennis Tito of the U.S. and Mark Shuttleworth of South Africa, to fly into space with the Russian Soyuz program.

Reading the article was one of those "wow" moments – like the time I saw a guy crawling out of the ocean with a scuba tank on his back and wanted to be able to do that so badly I could taste it. (I took up scuba diving right away.)

Hey…could somebody like me actually go into space? The requirements were pretty stringent. You had to be physically fit (I've always kept in shape), pass demanding medical tests (more of a problem, as we shall see), and spend several months in a "daunting training program."

There was a hefty price tag, too: $20 million. That's a lot of cash. But it was not a problem for me. Three years before, in 2000, my partners and I had sold my second company, Sensors Unlimited, for about $600 million in stock. We had founded the company on a relative shoestring in 1991. With my share of the proceeds, I had the money.

Incidentally, we bought the company back for $6 million – a penny on the dollar – in 2002, just two years after selling it. This makes for a very interesting story, to be covered at length a little later. It also makes for positive proof that a career in science can give you the ability to pursue your dreams.

In 2003 we were in the middle of the process of rebuilding Sensors. But I had been in the business for 11 years, and it was time to do something different. The idea of space travel was exhilarating, and I resolved to contact Space Adventures through their web site as soon as I got to work that morning.

On the Internet I found out that Space Adventures was (and still is) the only non-governmental organization that arranges orbital trips into space for private citizens. After some e-mails back and forth its president, Eric Anderson, showed up at my door with his vice president of orbital flight, Chris Faranetta. It was July 10.

A 28-year-old entrepreneur, Eric had founded Space Adventures about five years earlier. I could tell just by looking at him that this was a guy who didn't take no for an answer. (Not that he was likely to get a "no" from me.)

He had that drive that is so rare, but so necessary, to succeed in a startup business. He just reeked of enthusiasm. Eric was also an alum of the University of Virginia, just like me, with a degree in aerospace engineering. We hit it off immediately.

It amazes me how many people think that all you need to make a successful startup company is a fancy business plan and lots of money. That's 10% of it. The rest is hard work and relentless drive – never giving up!

To add to the coincidences, his sidekick Chris grew up in Rocky Hill, NJ, about four miles from where I was living. Chris and I also had a lot in common.

If you're a salesperson, these are the types of connections you like to have. They get the conversation going right away.

Although Eric is a great raconteur and salesman, it was pretty obvious he didn't have to push too hard on me. He described all the prerequisites for the flight, including how I would have to pass a series of medical and physical tests before being admitted to the program. Nothing fazed me – I was convinced that this was for me!

The lure of space

You might wonder why I was so instantly enthusiastic about climbing into a capsule and blasting off for the space station with its cramped quarters, bad food, and exotic methods of personal hygiene. There are a couple of reasons.

First, of course, I love a personal challenge. Orbital space flight was completely new, something I'd never done before, something very few people had done before. It was a great opportunity, and fortunately, I had the means to do it.

But there was also the motivating factor of curiosity about what it's like up there. I'm convinced that this curiosity is shared by most of the people on our planet.

As human beings we've always had this desire to know more about the world around us – and the one above us. Flying was a big part of that dream.

For most of human history we've wondered what it would be like to soar into the sky. The Greeks created the Daedalus myth, about an inventor who developed wings made of feathers to carry him and his son across the ocean. People built homemade flying contraptions and threw themselves off bridges and rooftops hoping to fly. Leonardo drew up plans for elaborate but impractical flying machines.

Finally, in 1903, the Wright brothers launched the aviation age. We had wings that worked.

Sputnik 1

Just half a century later something astonishing happened.

Sputnik 1, the first man-made satellite to orbit the earth, was launched on October 4, 1957. This was only 30 years after Charles Lindbergh had stunned the world with the first transatlantic solo flight. Now we were in space.

To make it more amazing, Sputnik was a creation of the Soviet Union. A team of Soviet scientists led by Sergei Korolev, the same man who later spearheaded the Soyuz program, had beaten the United States into space.

This wasn't just an empty metal ball floating high above the earth, either. Sputnik was small – only 23 inches across – but it had a sophisticated scientific mission. The satellite's radio transmitted measurements of on-board temperature and pressure. Analysis of the signal yielded data on radio-wave distribution in the ionosphere. Its orbits helped determine the density of the higher layers of the atmosphere. Any changes in these measurements could alert observers to meteoroid penetration of its outer hull.

Before Sputnik, the average American typically thought the Soviets were a bunch of barbarians – and Communists to boot! Our press and government helped foster this image. Maybe the Russkis were good at fighting wars and making vodka, but they were certainly not the intel-

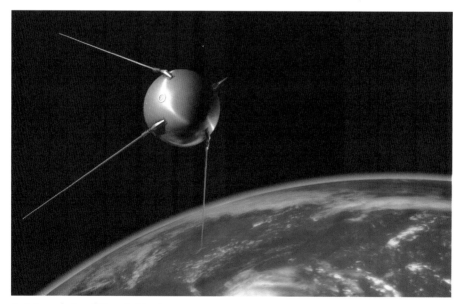

Artist's rendering of Sputnik 1 in space.

lectual equals of the good ol' USA! We had invented the light bulb, the phonograph, the airplane, mass production, the nuclear bomb. We were out in front to stay.

Well, they sure showed us. Their space program was actually ahead of ours! To further rattle our complacency, two years later their Luna 2 space probe hit the moon, the first human presence on earth's natural satellite. And in 1961 cosmonaut Yuri Gagarin became the first human to orbit earth.

What a gigantic favor they did us. I never fail both to congratulate and thank my many Russian friends for Sputnik, for Yuri Gagarin, the world's first astronaut, and for how they generally scared the shit out of us about our scientific competence. The shock woke us up. The Space Race was on, and we were determined to win it. How I wish we could get that same kind of wakeup call on our energy situation!

I was in seventh grade, and I still recall my teachers exhorting us students to learn math and science...to become engineers...to design rockets...to get to the moon first. On a national level the country rushed to revamp our educational programs, and within a year Congress passed the National Defense Education Act (NDEA), which funded everything from school construction to college scholarships.

Yes, it was panic time. But it was a good time – undoubtedly the best ever for American science. Billions of dollars poured into universities and centers of excellence in materials science, electronics, computers, chemistry, and other disciplines.

It worked. In 1969 we put the first people on the moon, and the Space Race was effectively over. As I go around now giving space talks to kids in hopes of getting more young people into science and engineering, I wish we had something like the moon race of 40 years ago to spur them on.

Party to negotiations

Eric Anderson proved to be as much of a go-getter as he appeared, and things moved fast. During our "negotiation" phase, Eric invited me to a party at Dennis Tito's house in Pacific Palisades, CA. This was in July, about a month after our initial meeting.

As the first space "tourist" in 2001 Dennis had blazed the trail for everyone who followed. It wasn't easy for him. Top officials at NASA opposed his flight, and according to reports in the *Los Angeles Times*,

NASA head Daniel S. Goldin even suggested that it was "un-American" for him to use the Russian Soyuz to go into space. He conveniently failed to mention NASA's own "un-American" partnership with the Russians. The two countries were cooperating to build the Space Station.

The official NASA resistance to Dennis got so intense that the day he and two cosmonauts arrived for training on the American sections of the Station, they were sent home by a NASA manager. The guy actually said, "We will not be able to begin training, because we are not willing to train with Dennis Tito."

That's bureaucracy at its worst – focusing on the petty stuff and missing the big picture. Fortunately most of the public, including NASA's own astronauts, supported Dennis. So did the media in the U.S. and Russia.

He flew on Soyuz TM-32 in 2001. He has testified before US House and Senate committees on the commercial opportunities in space. That's the mark of a man with a mission.

Serious fun

There were about 50 people at the party in Dennis's huge, gorgeous house. They included a number of Space Adventures people, both staff and consultants. NASA astronaut Buzz Aldrin, the pilot of the Apollo II lunar lander, was there too.

In addition to providing a good time with a spectacular view of the Los Angeles hills, the party was designed to expose potential spaceflight participants like me to people who had already flown in space. It went well – clearly they all had me sized up as a possible candidate, and no doubt Eric had Googled me to ensure I could afford the trip.

I wasn't there alone. I had felt a little nervous about showing up at Tito's palatial mansion overlooking the Pacific without seeing at least one familiar face in the crowd, so I invited along my good buddy, Rich Capalbo, who lived in nearby Pasadena.

At this point Rich was the only person I'd talked with about the space flight. For the most part I kept my plans quiet, not even telling my closest associates. I didn't want to tip my hand too early in the game, before everything was set, and then have to tell them I wasn't going after all. I'm the type who doesn't put the cart before the horse, especially when it comes to my emotions and expectations.

In that way I guess I'm a pretty buttoned-up guy – just ask my daughters and past girlfriends! But with Rich, things were different. We could tell each other anything, because we'd grown up together in Ridgefield Park, NJ. We'd done everything together, from playing little league baseball, to double-dating, to selling a company.

Rich had been a "financial consultant" for Sensors Unlimited during the Finisar buyout. We had drifted in and out of contact over the years, but had become close again during and after the Finisar transaction, including the buyback in 2002.

Funny, in our younger days, we would do anything together that involved making money: raking leaves, newspaper drives, loading trucks, shoveling snow. I guess that's why we both achieved success in business: we just never knew when to give up! We had probably been "dissed" by every employer who ever watched us scramble to earn a few bucks. But we never cared what people thought of us…as long as the money was green!

So when I was invited to the Tito party, I figured, why not ask Rich to tag along? He only lives a few miles away. When I told Rich what I was thinking about, he just laughed: "Space? Are you kidding?" When I said I wasn't and that we both had a chance to meet Buzz Aldrin, he readily agreed to join me.

Buzz lived up to his star billing. He greeted me when I arrived and regaled me with space stories. I was all ears.

I vividly recalled the Apollo 11 lunar landing and Neil Armstrong's famous "One small step" statement. It was a thrill to stand there talking with the pilot of the lunar landing module on that mission. Buzz was actually the second man to step onto the surface of the moon – he climbed down the ladder after Armstrong. I never imagined that I would follow in the foot-

Meeting Buzz Aldrin.

steps of these pioneers, but their achievement must have helped plant some seeds in my mind way back then.

It turned out that Buzz – like me – was from New Jersey (Montclair). It was amazing how I kept finding common ground with all the people involved in space endeavors. After serving in Korea as a fighter pilot, he earned a PhD in astronautics from MIT, so we also shared a science background. Buzz was on the advisory board for Space Adventures.

Buzz Aldrin is a quality guy, a real American hero. In spite of all his fame, he is an extremely gracious man, always making time for photos and a chat, always trying to motivate American kids.

Eric introduced me to the crowd of space folks whom I've come to know and respect over the years. Norm Thagard, who was the first NASA astronaut to fly on a Soyuz rocket, was there. He is a medical doctor who also holds a master's degree in engineering science.

There was a beautiful sunset over the Pacific. This was hard not to like. I think I knew clearly at that moment that I was going to pursue my shot at space travel.

Witnessing my future

If you haven't guessed by now, I'm not the type of person who ponders, considers, evaluates and waits – I usually just charge ahead! In spite of my scientific background, I'm more of a doer than a thinker.

At this point things were moving fast enough even for me. A series of discussions and contract drafts ensued, and in October of 2003 I found myself on the way to Russia with Eric and Norm Thagard.

We were going to witness a launch of the Soyuz rocket in Kazakhstan. Then I was to undergo a weeks' worth of medical tests to certify my fitness for flight.

We landed in Moscow, of course, and went from there to Baikonur, Kazakhstan. The trip is about 3 ½ hours. You won't find regularly scheduled flights to Baikonur – it's not a popular destination! You get there by private jet. Space Adventures had chartered a plane for us with about 50 people on board.

I remember asking Norm what it was like to train in Russia, and if he had any advice for me. He was the first American to fly to Mir, the Russian space station, in 1995. Norm's advice: "Just shut up and do what they tell you to do…you'll be OK." Good advice, as it turned out.

Later we discussed electronics and he wowed me with his knowledge of tube vs transistor audio amplifiers, regaling me with the finer points about odd and even harmonics. How much information can one guy keep in his head?

During the trip I got to meet Mark Shuttleworth, the second private citizen to go into space after Dennis Tito. It's amazing how these common denominators kept popping up between me and the space travelers I met. Mark is South African, and I had done post-doctoral work in Port Elizabeth, South Africa, in 1971-72. While I was there I fell in love with the country, and several years before this had bought a vineyard there.

After arrival in Baikonur we were escorted to the "Sputnik" Hotel to refresh ourselves. We didn't get much sleep, because we were going to a press conference at 3 a.m. to see the flight crew being interviewed before the launch. The launch itself was scheduled for around 9:30 a.m.

Two local Space Adventures people, Sergey Kostenko and Marsel Gubaydullin, escorted me through the process of getting to the launch. They gave me a first-hand education on how things get done in Russia. These guys really understood how to navigate the complex Russian space program, how to deal with the people, and how to get what you need.

It was from them that I learned that nothing is ever certain in Russia, right up to the very end. While this can be nerve-wracking, it does have an advantage. What you believe is impossible may actually turn out to be doable, often for reasons you never would have anticipated. Getting permission to take the scenic route to the viewing site for the launch was an example of this process, and of their skill in manipulating it.

The viewing stand at Baikonur is less than a mile from the launch pad. Keep in mind that the propellant for Soyuz is jet fuel mixed with liquid oxygen. Its combustion byproducts are reasonably benign.

NASA, on the other hand, keeps you a couple of miles away from the launch pad at the Kennedy Space Center. They do this because their fuels are toxic, and the blastoff leaves a bit of hydrochloric acid in the wind that eventually might rain down upon you.

Finally, we arrived at the viewing site. We would be close-up witnesses to 20 million peak horsepower unleashed right before our eyes!

It took just another 20 minutes for the launch to occur. At that moment I saw a brief orange flame shoot out of the rocket, followed by the deepest bellowing roar I had ever heard in my life. Even though it was a bright morning, the sky was lit up with a far more brilliant light, and the ground shook like we were at the epicenter of a magnitude 8 earthquake.

Slowly at first, then with increasing acceleration, the rocket rose into the sky and disappeared. It was such an awesome sight!

My guides had warned me to bring welding glasses and earplugs, but I had neither. I was wearing nothing more than ear muffs to protect against the 40-degree temperature and wind, plus sunglasses to reduce the natural glare.

After the launch my ears felt like I had been front row center at a rock concert for ten hours. I was hearing a dull roar for the next day. My eyes took several hours to recover as well. You have to listen to the people with experience if you don't want to get into trouble.

There was applause from the viewing stand following liftoff, then relative silence for the next seven minutes. The viewing stand got a play-by-play account of the launch from the public address system. In the control room the staff could watch video of the crew inside the Soyuz as it climbed into orbit. There are three cameras inside the ship, pointed at the mission commander, the flight engineer, and the third seat where I would be if I made it through medical tests and training.

About 9 minutes after blastoff, the loudspeaker announced in Russian that orbit had been achieved. Applause erupted from the onlookers, vodka and cognac bottles emerged, toasts followed, everyone celebrated.

When the celebrations were over, we trudged back to the hotel, got maybe an hour's sleep, then packed and headed back to the airport for our return to Moscow. I had a week of medical exams ahead of me.

Medical blacklist

Before the doctors got started, I was scheduled to take a ride in a Russian MiG 25 high-altitude interceptor plane so I could see what life was like at 80,000 feet and more than twice the speed of sound. So off I went to the Zhukovsky Flight Research Institute, about 20 miles southeast of Moscow.

During the pre-flight I was checked by a kindly, gray-haired, grandfatherly-looking man in a white lab coat, evidently the medical examiner. He asked me some basic medical questions which I answered positively. He looked up, smiled, and asked "Are you ready to fly?"

Of course I said yes. To my surprise, he proceeded to take off his lab coat and put on a flight jacket. He was also the pilot! We walked out to the runway.

Since this was Russia in October, there was snow on the ground, and a truck driver was busy plowing it off the runway. My pilot motioned to him to come over and tow our plane out for takeoff. Grumbling to himself, the plow driver hooked his plow to the MiG, gave us a tow, and went back to his regular job of plowing snow.

On an American airstrip there would have been a special aircraft tow tractor sitting there, completely idle, waiting for the next plane that needed a pull. In Russia the tow tractor doubles as a snowplow. The Russians make maximum use of limited resources.

My pilot sat behind me – I was in the front compartment of the cockpit – and explained how to eject from the plane if we encountered a problem. It was a bizarre version of those pre-flight emergency procedure lectures you get from the cabin attendants on a commercial airliner. Instead of cautioning me to stay quietly in my seat in case of trouble, he was telling me to pull up on a very stiff lever as hard as I could to launch myself out of the plane!

Then we were off. The MiG roared into the air, accelerating at a fantastic rate, until we reached MACH 2.5, two and a half times the speed of sound. Soon we were at 80,000 feet.

It was thrilling. At peak altitude I could see the curvature of the earth below me. Above there was only the blackness of space. I was wearing a pressure suit and oxygen mask, so it was a preview of what I would experience in space flight. There was even a fleeting sense of weightlessness as we came down.

Before takeoff I'd wondered if I would like it. After that 45-minute flight I was exhilarated.

Back on the ground, the snow plow came over and pulled us back to reality. It was time to see the doctors.

Medical uncertainty

My first medical exam was in the main building of the infamous "IBMP," or Institute of Biomedical Problems. It's an apt name, because they are definitely focused on finding problems. Its testing facilities are scattered across about a dozen different hospitals throughout Moscow and surrounding areas.

IBMP is tasked by the Russian Space Agency with filtering out any cosmonauts who have medical conditions. These guys are charged with medically certifying anyone who flies on a Russian space vehicle, and they take the job seriously. They're very conservative and quick to point out even the most minor health issues.

In their defense, imagine what might happen if they approved someone who then developed a serious medical condition up on the ISS. It would jeopardize the whole mission. More seriously, contagious diseases can endanger the entire crew. IBMP doctors have to be tough.

IBMP is "old school," not only in staff age, but in technique. While the individual doctors are highly competent, they don't always have first-class medical facilities or diagnostic equipment. It's another instance of the lack of infrastructure I was to encounter repeatedly in Russia.

That first exam, for example, consisted of the usual blood tests, urine samples, and pulse/EKG measurements. To collect urine samples they gave me pint-sized glass bottles that, as far as I could tell, were washed and reused. They seemed to use very little computer-controlled equipment or record-keeping.

There were also a lot of "hands-on" exams. Russian doctors seem to examine the body manually more than American doctors do, often without rubber gloves. I don't know if this is good or bad, just different from what I was used to. Overall it was eye-opening to see the level of medical care available even to upper-income people like me, not to mention the cosmonauts.

In my case there was an "elephant in the room" medical issue. I had suffered a collapsed lung in 1999. More specifically, a spontaneously collapsed lung or, as the docs say, a "spontaneous pneumothorax." This means that the lung did not collapse from a trauma (auto accident, impact to the chest, etc.), but literally deflated all by itself.

Now, I never hid this, and as I recall my first phone conversation with Eric Anderson went something like "Hi…I'm Greg Olsen and I want to fly into space and I've had a collapsed lung…."

I knew it would be an issue and wanted to be up front about it. As it turns out, I had no choice. The first part of the Russian medical exam is to receive a detailed medical history from your American doctor, which would include this kind of incident.

There is usually an underlying reason for a spontaneous pneumothorax. In my case it is hereditary. My Dad had a collapsed lung, my sister had two, my niece one.

It seems that people in our family are born with "blebs" or "bubbles" in their lungs, much the same as the series of small bubbles you sometimes see in cheap glassware. Blebs can range in size from millimeters to inches long, but they usually have no direct effect on your health or breathing.

No effect, that is, until one of the blebs decides to burst. This allows the air to leak out from your lung into the chest cavity, effectively neutralizing that lung. The collapse can be caused by coughing, sudden air pressure changes (like you undergo in airplanes or, for that matter, space flight), a gradual weakening of the wall of the lung, etc.

Smoking and blebs

In my case that's where behavioral factors come into the picture. Everyone in my family who had a collapsed lung was also a heavy smoker. We really abused our lungs with cigarettes.

I'd smoked for many years. And I mean many – I'd started at the age of 6, and continued through my high school years. By college I was a doing a steady 1½ packs a day. It started with Lucky Strikes, then went to Kools with menthol, which were supposed to be "soothing" or somehow better for you. People actually believed that kind of baloney in the 1950's!

That was an era when twice as many people smoked as they do today. In my days as a researcher (1972-73) everyone smoked around delicate equipment, like electron microscopes, and thought nothing of it. We even smoked while using flammable chemicals!

Every fall and winter I seemed to get bad coughs. In September 1973 I went to a doctor in Trenton, asking for relief from my hacking cough. He just glanced at the Kools in my shirt pocket, smirked, and said, "Keep that up and you'll be back here every year."

The pain in my lungs probably had more to do with my smoking habit than his sarcasm, but his words lingered in my mind. I decided then and there: NO MORE CIGARETTES! And I've not had one in over 35 years.

I used to have an occasional cigar, but cut that out too about 15 years ago. Don't let anyone fool you: cigars can be even worse than cigarettes.

Even though I'd quit, some damage had already been done. Combined with my blebs it made me a prime candidate for a pneumothorax.

You can get a collapsed lung without being aware of it, because if your other lung is still functioning, you can still go at about half-speed. My first inkling that something was wrong came during a technical conference in Davos, Switzerland in late May of 1999. I was there with coworkers Mike Lange and Chris Dries. We decided to go on a hike up into the mountains and I found myself lagging behind both guys.

They are both much younger than I, but I pride myself on being reasonably fit. I was surprised at how "out of shape" I was. I didn't pay it much mind after my return but did seem to notice "cold symptoms," including a persistent cough.

A few weeks later, I figured I should visit my doctor, get some pills and shake this thing. When I went to see him, he examined me and gave me prescriptions, and just as I was walking out the door, said, "You know, maybe we should get a chest x-ray just to be sure. Here's a prescription for that too. See if you can get it done in the next day or so."

I said OK, and since the hospital is right next to his office, asked if

I could get it done right then. He said it was after 5 p.m., when they stopped doing routine scans, but he had a friend in the x-ray department and would make a phone call.

I had a 6 PM dinner engagement, but figured I could squeeze it in. Sure I could!

The next thing I know I'm lying on a hospital bed with a chest tube in my side. I had been walking around with a totally useless lung for several weeks. It was just another reminder of how life is what really happens in spite of all of that planning we do.

I spent the next two weeks in the hospital and received a procedure called a "pleurodesis." This is something like "fix-a-flat" to help keep your lung from deflating again like a blown-out tire.

One lung was fixed. But the other lung was still subject to the whims of the blebs, as the Russian doctors knew.

Never give up

After a week of being chauffeured around to numerous medical buildings in Moscow, the time came for my final medical review. It was not good.

The IBMP doctors seized upon the collapsed lung and the possibility of the other one collapsing in space as reasons to question my fitness. Even though normal space travel does not include major pressure changes, "off-nominal events" – a wonderful phrase to make looming disasters such as cabin leaks sound like minor technical problems – would expose me to reduced pressures that could affect my blebs.

They also worried about the increased pressure on the chest cavity due to the elevated "g-forces" during launch and re-entry.

To my surprise, the medical guys quibbled with some heart measurements as well. My annual physical usually ends with a slap on the back from my doctor, praising my low cholesterol, low pulse rate and blood pressure, and good EKGs.

But the Russian docs had found some occasional premature ventricular contractions, which are slightly irregular heartbeats. You will find these in most people over 50 (many younger as well) but, as I've mentioned, Russian doctors take their job very seriously. They basically disqualified me from the program.

Feeling rather depressed, I went back home. I was glad I had told so few people of my plans. I could just picture dozens of well-meaning friends coming up to me and saying, "…hey, Greg, how did you make out in Russia…are you gonna go into space?"

The truth was, I didn't know, and things didn't look all that good. The Russian docs really sounded pessimistic. "You should see your doctor in the U.S. immediately," they said, shaking their heads

It was all very unsettling. I get annual physicals and thought everything was fine. Outside of my lung, there were never any medical issues to be concerned about. I've always prided myself on being healthy and fit.

In fact I was ready to live forever, which is one of the things that people think is "off-the-wall" about me. I have this faith that I am going to live for a long time. And why not? Life on earth has been pretty good to me. I don't like to dwell on the alternative.

But now I had to think in terms of mortality. I did revisit my doc for my annual physical, and things were fine as usual. This was encouraging, but carried little weight with the Russian medical establishment. It was time to consider my options.

One person I had told of my plans to go into space was my daughter Krista. She had been living in my house with her husband and three kids while they were planning and building a 15,000 sq ft "McMansion" in Colt's Neck, NJ. So when I decided to go to Russia, I told her why.

As I recall, when she heard what I had in mind, she laughed and said something like "That's just the sort of thing I would expect from you, Dad." But I got no objections or fearful concerns, from her or from her sister Kimberly.

If my daughters had confidence in my fitness to fly, and I was sure I was up to the task, why back down now?

Eric Anderson enlisted Dr. Richard Jennings, a consultant for Space Adventures, to try to persuade the Russians to reconsider. He is a former Chief Flight Surgeon for the Johnson Space Center, and runs a flight surgeon residency in the Aerospace Medicine Center at the University of Texas Medical Branch (UTMB).I met him at his office at UTMB in Galveston.

Dr. Jennings is a down-to-earth Oklahoman, very practical. He also is a pilot and knows firsthand the issues related to medicine and flight. I remember him saying, "How do we get you from here to flying?" He said his philosophy was that if a pilot had a minor heart ailment, rather than put him on medication, in many cases he would have that guy spend an extra hour in flight training.

According to him, more accidents are caused by human error than by medical problems. Almost everyone has some minor medical condition, but it's rare for somebody to have a heart attack, for example, and crash a plane. Accidents usually happen when a recreational pilot in a private plane gets in over his or her head and becomes disoriented.

Dr. Jennings was a strong advocate for me. He researched lung problems and came up with a possible solution, which he shared with me. It turns out that commercial pilots, who are typically tall, thin, athletic men like me, are often prone to spontaneous pneumothorax (lung collapse). The rule of thumb for them is that if they don't have another collapsed lung for five years, they can fly again.

But pilots who elect to have a pleurodesis of the uncollapsed lung – which involves a deliberate, surgically induced collapse – can fly commercially without the waiting period. On this basis Eric Anderson started further discussions and negotiations with the Russian side.

It was like running around the Baikonur launch site with Sergey and Marsel all over again. Obstacles? There were plenty, but there's always a way around them, if you just look long enough and argue hard enough. Eric Anderson has become a master at this, out of sheer necessity. As Rich Jennings said, almost every civilian space candidate will have some medical issue that could potentially prevent him or her from flying.

We ended up working out a compromise with the Russian doctors, who agreed that if I had a pleurodesis performed on my other lung, they would tentatively admit me to the training program in April of 2004.

How badly did I want to get into that program? Here's how badly. I was elated at the prospect of undergoing an hours-long surgical procedure that involved anesthesia, chest incisions, tubes and video cameras inserted into my chest cavity, not to mention weeks of recovery. It was positively masochistic.

Boy … when I want something, I'll go to the ends of the earth to get it!

Surgery and aftermath

Fortunately there was no need to travel that far. The renowned Deborah Heart and Lung Institute in Browns Mills, New Jersey is less than an hour's drive from Princeton.

How I came to have my elective pleurodesis performed at Deborah is the result of one of those interesting, random events that seem to have shaped my life. Back in the early 90s I had met a woman at a University of Virginia alumni event and we began dating. Years later, even though we were no longer dating, her sister contacted me by e-mail as part of a fundraising event for Deborah. She'd probably seen all the then-current publicity on the profitable sale of my second company Sensors Unlimited.

The sister set up a dinner at Rat's restaurant in Hamilton, New Jersey (despite its name, Rat's is a four-star eatery) so I could meet Dr. David Murphy, chair of Deborah's pulmonology department at Deborah. Dr. Murphy was looking for funding for an experiment to go into the chest cavities of people with collapsed lungs and image the rupture of the lung.

Wow, I thought, that's just the sort of thing I'm interested in, too: collapsed lungs and imaging. We hit it off, he sent me a proposal, and I agreed to help fund his research project.

I had gone to a couple of other lung doctors after the spontaneous lung collapse and pleurodesis, but wasn't really happy with any of them. They just told me what I already knew, gave me some inhalers, and sent me on my way. Dr. Murphy was different, a real scientist, so I adopted him as my pulmonologist.

When I came back from Moscow, I went right to Dr. Murphy and asked him to review what the Russians were saying about my health. He said that even though my lung had once collapsed and I had a mild case of breathing obstruction (chronic obstructive pulmonary disorder, or COPD), there was no reason why I could not live a normal, healthy life.

I agreed. I can dance for three hours, swim the length of a pool underwater, and run for miles. I go to the gym and do high-level aerobics. That's normal and then some.

Dr. Murphy arranged for me to have the voluntary pleurodesis done at Deborah in early February. Recovery from one of these surgeries is not trivial. First the lung has to reinflate, which took almost a week in my case. Then there's the rehab, which is largely a matter of attitude.

It's funny who you trust in situations like these. I wound up turning to Joe Stach for some moral support.

Joe and I had been casual acquaintances for several years. We served as best men when our mutual friend Lubek Jastrbski remarried. But eventually, thanks to serving together on the boards of Lubek's company and one other firm, we got to know each other better.

We were roughly the same age and had similar backgrounds. We could talk shop for hours, and just as quickly switch to lightly discussing Lubek's latest antics. While I was recuperating at Deborah Joe called me, I think to discuss some business. For some reason – maybe because he lived close by – I invited him to the hospital.

I doubt more than a few of my closest friends knew what I was doing, but somehow I wanted to confide in Joe. He is one of these rare people who is both very competent technically, and has good insight into human nature.

We shared some laughs about the lengths I was going to in order to reach space. It helped a lot to get his support.

But an even bigger boost came when Krista brought my three young grandchildren in to see me one night, just before I was about to go to sleep. That really made my day!

There were a couple of aftereffects. I had a weird experience shortly after getting out of the hospital. I am no malingerer – as soon as humanly possible after an illness or injury, I'm up and on my feet. A good friend had driven me home from Deborah, and the very next morning I jumped into my car to go to Home Depot or someplace.

There's a blind corner about a mile from my house. I stopped at the stop sign, and the next thing I remember is brakes screeching, a horn honking, and a guy who had barely missed plowing into me running off the road into the woods. To this day, I have no memory of how that accident happened.

I am not the world's greatest driver. It's not that I'm reckless, but sometimes I can be thoughtless – and I usually know when I've made a mistake, like cutting a guy off. But this time: not a clue. They say that the effects of anesthesia can hang around for a week or more. Maybe that's what happened to me.

Fortunately, no one was hurt, and there wasn't even any body damage. I did have to pay to get the guy's car towed, and he even insisted I give him money to get it washed, which I gladly did.

That was pretty scary. For the next several days I was much more careful, though I gradually reverted to my air-headed self.

About a month later, while staying at a hotel in Washington, DC, I decided to jog a few laps around the block to stay in shape. Once again it became clear how serious the operation really was. About halfway through the first lap I suddenly felt totally exhausted, and had a dull pain in my chest.

"Oh well," some might think, "I'm 58 years old – I've just had surgery – better take it easy the rest of my life…." But not me. My masochism didn't stop with volunteering for surgery. I wasn't going to let a little pain stop me from running. And besides, I was sure running would be a required part of my training in Russia. Better get ready!

Next day I made it around three-quarters of the block. Just as people have to relearn to walk after an accident, I suppose you have to learn how to breathe hard again after lung surgery. So I kept at it, and even though I'm no big fan of jogging, I was soon up to my self-imposed goal of two miles a day.

I was making progress, and space was looking just a bit closer than it had since October. As my recovery quickened, my optimism picked up as well. Encouraging murmurs were emerging from Space Adventures.

Now all I had to do was get back to Russia and start training.

Official space portrait of (l to r): Gregory Olsen, Cosmonaut Valeri Tokarev, and NASA Astronaut William McArthur.

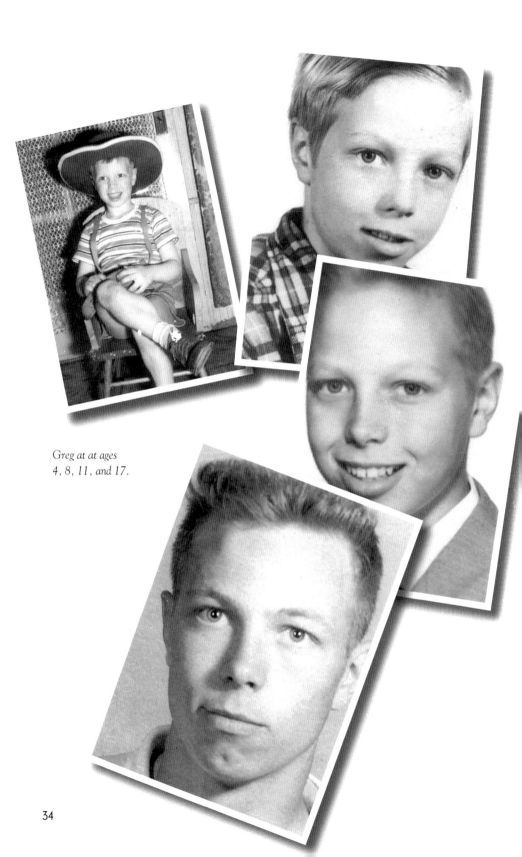

Greg at at ages
4, 8, 11, and 17.

3 Flight Prep: Early Days

So there I was, waiting around in Moscow while a team of Russian doctors decided whether or not I was medically qualified to go into space. How did that happen? There certainly wasn't much in my background or my early years to indicate I'd ever get even this close to something as special as going into space.

When you understand where I came from and who I was, you have to conclude that anyone has a chance to live their dreams. Most people who knew me as a kid wouldn't have tabbed me as someone destined for success.

But it does seem like I was always trying to get ahead, even as a preschooler. I was admitted to Kindergarten when I was only 4 ½ years old. That would be a stretch for most kids, but it really wasn't that tough for me. I could already read, write and do simple arithmetic, thanks to my schoolteacher Mom.

Being ahead of the average kid helped, but it wasn't enough to get me into school at that age. That's where Mom came in. As a schoolteacher she knew people in the system, and could find a way for me to jump the queue. As a result, I got a year's head start in life!

We were living in Bay Ridge, Brooklyn, at the time. I was born there in 1945, and spent my first two years of school at PS 185 in my neighborhood. Those were fun times – and the fun wasn't all in the classroom. Not only was I ahead of the other kids in classroom work, I was way out in front in extracurricular mischief-making.

I remember being on the school lunch line in kindergarten, with only a nickel in my pocket, and asking the kid in front of me to trade his dime for my nickel. I reasoned with him that my coin was better because it was bigger. He actually bought it and made the trade!

Conning a fellow student was one thing. Cheating a grownup was another. At about the same time one of my classmates taught me how to get on the 86th St bus without a pass. Talk about having the wrong kind of mentor!

He showed the driver his pass, got on the bus, then walked to the back door and flipped it out to me. I flashed his pass at the unsuspecting driver, who let this innocent-looking little blond-haired, blue-eyed boy get on without really looking at his ID.

We also used to hang out in vacant buildings, especially those under construction. I remember this watchman on one job who let us have sodas on credit, on the promise we'd pay him back some day soon. He was probably some type of pervert, but we thought he was Santa Claus.

Finally one day he demanded his money. When we couldn't pay he began shouting at us, and chased us down the block. It was an early lesson in the old truism: there is no free lunch!

One of my last memories of Brooklyn was when another friend and I decided to start a fire in a vacant lot, just to watch it burn. I guess we were pretty good at arson, because three fire trucks showed up to put out the blaze. Those sirens scared the hell out of me!

That incident taught me another important lesson – this time a negative one. Sometimes bad behavior gets more attention than it deserves.

Think about it – it happens all the time. Teachers pay more attention to troublemakers than they do to kids who behave well. Within a company the boss seems to spend an inordinate amount of time dealing with office malcontents, while neglecting the worker bees who are actually getting the job done.

The pattern extends to venture investments, too. When startup companies I've invested in are doing great, I often treat them with "benign neglect," and let them do their thing. Would they do even better if I was more involved? I don't know. But I do know that I spend far too much energy on the whiners and complainers. The stories are always the same: "….we don't have enough capital….salaries are too low…we failed not because of me but _____." Instead of focusing on the good people, I find myself getting stuck in a negative situation dealing with a bunch of losers.

Maybe that was how the adults around me felt about young Greg Olsen. I was a kid whose primary mode of attracting attention was getting into trouble. It may have been the only way to get attention in my family, but whatever the reason, I did it all the time, within the family and outside of it.

Acting up in the Poconos

The pattern continued when we moved out of Brooklyn. Dad had worked in the Brooklyn Navy Yard during World War II as an electrician and welder, where he got friendly with a guy named Connie Cortright. He and Connie used to go hunting in the Pocono Mountain region of Pennsylvania, where some of Connie's family already lived.

One day in 1951 the two buddies got the brilliant idea of making their fortunes by buying a gas station in the Poconos, in Rowland, PA, which had a population of about 50. During the summer maybe a couple of hundred people would drive up from New York City for a vacation. But come September, the potential customers dropped back to 50. And most of these people hardly had enough money to buy gas, let alone get their autos repaired!

My Dad's take from the gas station was maybe $20 a week. Even in 1951 it was tough to make a go of it on that income. So to make ends meet, he went back to his old electrician's job in New Jersey. He stayed there during the week, leaving his wife and four kids (ages 1, 3, 6 and 9) all alone in this rural backwater.

How backwards was it? Well, we had an outhouse for a toilet, a hand-pumped well for water (which often froze in the winter), and a wood-burning pot-bellied stove for heat. I remember my Dad going down to the sawmill in late September, picking up a pile of side cuts (the pieces left over after they squared off tree trunks into lumber), and showing me how to chop them up to fit in that pot-bellied stove.

If someone tasked a six-year old kid with that kind of job today, they'd probably be hauled up before a state agency for child endangerment. But I loved it, and it was certainly not uncommon 50-odd years ago.

That was just one of the realities of life in the Poconos. These days we joke about how parents try to convince their kids that they have it easy: "When I was your age, I walked to school…through six feet of snow…up hill – both ways!" Well, it wasn't that bad where we lived, but we faced situations that today's kids never experience. For example, we had to walk about a mile in all kinds of weather to and from the school bus stop. Nobody thought anything of it.

School in the Poconos was a big change from Brooklyn. I was in a two-room schoolhouse, grades 1 to 8, in Lackawaxen. Each row of seats was a separate grade. On Friday mornings we had spelling tests. My teacher, Mrs Delert, would have four spelling books laid out on her arm, one for each of the grades in the room, and she would call out the words for each grade to write down.

We've all read the nostalgic tales about how wonderful the one-room rustic schools were. People imagine dedicated teachers encouraging eager country kids to learn, and upper-grade students helping the younger ones advance.

I hate to burst that bubble, but not everyone in my two-room country schoolhouse was all that bright or all that interested. They thought more about hunting and fishing than school. Even though I was only six, my Brooklyn education had put me so far ahead that I was skipped a grade. Combined with the early admission to kindergarten, that made me two years younger than anybody else at my grade level!

The age gap wasn't all that relevant, though, since the only other boys in town were some six years older than I. They weren't the greatest role models in the world, either. Following their example, I started smoking – at age six.

I also learned from them – willingly, I must admit – how to commit such petty larcenies as distracting the owner of the general store while others stuffed candy bars into their shirts; breaking into the store after hours and stealing a case of 8-oz Pepsi's; and filching cigarettes from my parents' ever-present packs.

Soon my pranks earned me my first brush with the law. I was hanging around with Ritchie Cortright, Connie's 12 year-old son. We went out and smashed a few windows on someone's house in Rowland – just to hear the glass break, I guess. No one was home, but eventually the owners discovered the damage and called the state police.

About a week later I saw this patrol car pull up to the school, and out stepped a state trooper. He must have been 6'5", with two huge pistols at his side, a Smokey the Bear hat, and a walk like John Wayne. I knew why he was there, and I was shaking all over.

It didn't take long for him to finish his discussion with the teacher, and I can still remember him looking me square in the eye, pointing his finger and waving me into the kitchen area.

"Now listen, kid, there were some windows broken last week in the Merrill home," he said. "Do you know anything about that?"

"Oh no…." I replied, but the fear on my face must have been evident. I didn't know if he would smack me around – common practice back then – or just yell at me.

He did neither. Instead, he just smiled, pulled out a Hershey bar, split it in two and offered me half. "Now look kid…I *know* you broke those windows. Just tell me how many you smashed, and how you did it."

His peace offering so disarmed me that I blurted out that I broke 4, maybe six windows.

"How?" he asked. "I threw some rocks," I said. "Threw rocks from where?" "Oh, maybe 200 feet," I guessed wildly.

Apparently they thought we might have broken into the house. Once I convinced him that we only threw stones from a safe distance, he seemed relieved and said "200 feet, huh?" He must have been amused at my wild estimate of the distance.

"Now listen, kid, if I ever catch you doing this again, you're going right in that police car and straight to jail," he admonished me. Then he motioned for me to leave the room. But he had a twinkle in his eye and gave me a big wink as I left.

Ten minutes later Ritchie went through the same interrogation, but as an older guy, he got more intimidation. I believe reform school was mentioned to him as a future destination. This was the standard threat made to every misbehaving boy in the 1950's.

Ritchie didn't seem very relieved when he walked out of that room. But later that afternoon, we were both laughing about it, and probably planning our next caper.

You'll notice that the trooper didn't ask if anyone else was with me. He had grown up in the area – he knew I wouldn't rat on a friend even if he threw me in the river. You just didn't do that.

But he knew who the other guy was anyway. Ritchie and I were the only kids within miles of that house. Who else could it have been?

Family matters

If rural northeastern Pennsylvania wasn't the idyllic setting of the storybooks, it was a good match for how our family approached life.

In my family, life was something you got through. You worked, you ate, you smoked, maybe you laughed it up with the guys at the bar or drowned your resentments in whiskey at home. As for happiness, who aspired to that? Life was an endurance test.

My parents weren't very demonstrative, and giving their kids pats on the back didn't come easily to them, especially my Dad. He was second-generation Norwegian, and as a result of his upbringing he displayed a Nordic reserve with his children.

In fact, neither of my parents paid us much attention, unless they thought we needed correcting. They never had a lot of money, and their life was focused on earning enough for the necessities. That didn't leave much time for coddling their kids.

One incident that typified this attitude sticks in my memory. Shortly after we moved to Rowland, my dad, in a rare moment of magnanimity, ordered me and my older sister Amy new bikes from the Sears Roebuck catalog.

In 1951 that was a very big deal. Having my own set of wheels would infinitely improve my mobility – and increase my social standing at school.

But Dad could be capricious, especially when it came to me. Here was a man who would eagerly buy the guys at the bar a round of drinks but deny me a new pair of sneakers when I sorely needed them. Besides, I was the family troublemaker. When I committed the inevitable minor infraction, Dad canceled my bike.

Neither Amy nor I can recall what I did to deserve this fate. Maybe I pulled one of my sisters' hair, or was clumsy and spilled milk at the dinner table. Whatever the cause, I took it pretty hard; I was really counting on that bike. Amy got hers and I got nothing. Naturally, when I wanted to ride, which was frequently, I stole hers, feeding the vicious cycle of misbehavior followed by punishment.

Eventually I got over my disappointment, as most kids do. It was one of the experiences that taught me what I came to adopt as a philosophy of life: adversity is a challenge, not a defeat.

I will say this about my Dad: he was a smart guy. He was not highly educated, but was very interested in learning (more about inanimate things than human relationships and his family, unfortunately).

He loved to read encyclopedias. In the 1950's door-to-door salespeople were still common. One day an Encyclopedia Britannica salesman rang our bell and my father wound up signing a contract for $700 to buy the whole set! He figured we kids would use it for school.

As it turned out, we rarely looked at it – it was too abstract and boring. But when Dad wasn't at the tavern, he could often be found in the living room reading some article about ancient Egypt, hydraulic dams, or whatever else interested him at the time.

He was very knowledgeable about electricity and prided himself on being a first-rate electrician. He actually explained the "three-phase" electrical system, in use throughout the world, to me when I was in high school. And that required some understanding of trigonometry!

Jersey days

In 1954, when I was nine years old, the family ended three years of backwoods exile and joined my Dad in Ridgefield Park, New Jersey. I lived there right through college.

There were four kids – three sisters and me –in a two-bedroom rented house. That made for a truly hectic daily routine. Although our mother occasionally did substitute teaching, she was basically a stay-at-home Mom.

School was different, too. Instead of a two-room schoolhouse I went to a real school with separate rooms for each grade. Since there were lots of kids in my grade who were smarter and more motivated than my classmates in the Poconos, the educational advantage I'd had vanished overnight. I did OK in grade school, but that was all.

In fact, the school put me back a grade, not because I couldn't do the work, but because I was always getting into trouble (of course). Mischief seemed to be one area where I was still a leader, even though I was still a year younger than my new classmates.

There was a bigger choice of extracurricular activities in small-town New Jersey than in the Poconos. Instead of hunting and fishing, I played Little League baseball. One of the guys in my league was Rich Capalbo, who years later provided financial advice when we were negotiating the buyout of Sensors Unlimited.

He was pretty savvy about money even then. A kid could actually earn money in Ridgefield Park. I started working for spending cash at nine years old, doing paper routes, lawn mowing, and other odd jobs. Rich and I were always involved in money-making schemes of one kind or another right up through college.

My later entrepreneurial bent may have been foreshadowed in those years.

Delinquent with diploma

Not all of my for-profit enterprises were so innocent. I wasn't above a little petty larceny – my mentors in the Poconos would have been proud. Like other misbehaving teens back then, I picked up some ready money by swiping hubcaps off cars.

Understand, these weren't the cheap practical kind of hubcaps. They were more like fashion statements. Popular styles included chromed half-moon models, ornate types with elaborate wire spokes or intricate designs, and ones with "spinners," propeller-like attachments that whirled around while the car was running. They could bring a good buck on the underground market, making them tempting targets for sticky-fingered teenagers.

But I guess I wasn't all that good at minor theft. The police caught me and my partners in crime red-handed, and in 1961, when I was 16, I was convicted of juvenile delinquency for my hubcap exploits. New Jersey cops were less willing to let boys be boys than that state trooper back in Rowland, PA.

By this time I was in high school and not doing all that well. Even though the Sputnik launch got me interested in science in 7th grade, I was struggling in math. Failing my trigonometry class in senior year was the low-water mark.

That's not to say I was just hanging out, doing nothing. I continued working part-time, and earned an amateur (ham) radio license on my own. But schoolwork got the minimum possible investment of time.

To top it off, I had what many people would call an attitude problem. As I've already mentioned, I was suspended from high school twice, and truthfully, probably deserved worse. I thought the way to win the approval of my peers was to stand out as a character who would do outrageous things. I probably spent half of the afternoons of my entire four years in detention with the other troublemakers, where we would do more mischief.

We had this one teacher who had a "twitch" problem with his eyes. We called him "Squint." One of us in detention would shout "Squint" when his back was turned. As he ran to find out who did it, another "Squint" would be uttered from the other end of the room. Detention time would be doubled, and re-doubled, but we didn't care. We were basically about getting laughs.

Our status slowly increased among our classmates as we gained a reputation for standing up to authority and generally having a good time. I did some things that I'm not proud of today – things that were actually quite dangerous. We were always looking for more inventive ways to interrupt the normal flow of the school to upset the teachers … and especially our hated principal, Mr. Arbo.

One Friday afternoon I got the brilliant idea to screw a penny into a light socket in the boys' locker room. I thought it might be a roundabout way to sneak into the girls' locker room, which was a perennial goal.

I knew that once I screwed that bulb back in, it would short out the contacts and blow a fuse. Well, it sure did. And it took them about 3 hours to figure it out. Gym classes were cancelled. I still remember Mr. Arbo's threatening voice on the PA system saying, "...whoever did this...I assure you we will catch you and we WILL expel you from school."

They never did figure it out. One of my buddies asked me "Hey Olsen...was that you who did it?"

"No, it wasn't me," I replied, but he could tell from the twinkle in my eye that it was!

My second suspension came during my senior year. We had another teacher we called "Meatball" because he was short and fat. We gave him the same treatment in detention as we gave to Squint, but Meatball was a little meaner when he caught wrongdoers. One evening about five of us decided to drive over to his house and see if we could find him.

It was in February, as I remember, and there was snow on the ground. He lived in a third-floor apartment, and when we shouted "Hey Meatball" in front of the building, he actually came to the window, opened it up, and stuck his head out! Snowballs flew through the air, and one actually hit him.

Next day, a parade of the members of our group was seen going in and out of Mr. Arbo's office. "I know you were there, Olsen," he growled at me with a threatening finger wag.

"Oh, no … I was home studying," I lied. But when I saw one of the guys (he wasn't in our main group) coming out of the office in tears, I knew it was curtains for us.

I continued to deny my guilt when my parents had to come to the office, but Mr. Arbo told my mother that I was nothing but a Philadelphia lawyer. That's when he guaranteed me that I would never get into any college in the United States.

I got suspended from school for a week. By some miracle, I was actually able to graduate, although the senior prom and any social activities were clearly prohibited for me. The suspension was handed down on the day of the prom. This was really bad timing. I had a date for the prom, and now I couldn't go because I'd been suspended. So my date would have to sit home too.

I truly felt terrible about having to stand her up. To make things right, my sister Amy and her boyfriend proposed a double-date for dinner at a restaurant in New York City that very night. In a way it was better than a prom. We got to play grownup, including having a drink

With prom date Nancy Herbeck, sister Amy, and Amy's boyfriend Bobby Sayers (l to r) at the Hawaiian Room in NYC, May 1961. I had been suspended from school that day and could not attend the official prom.

(New York's legal drinking age was 18, and we were close enough).

In spite of all the negatives in my high school career, I do have some fond memories. One big one involves athletics. That's pretty strange, since athletics were never a strong suit for me.

You have to know, first of all, that high school football is like a religion in New Jersey towns. People in Jersey are almost as crazed about the exploits of their local team as Texans. In my senior year I got to be part of the glory. I actually made the football team.

I didn't start, of course – I was tall and skinny, which is not a good body type for football unless you have a ton of talent, which I didn't. But I did play on special teams (kickoff returns and the like) and won my letter. This was one more grace, something I hadn't really earned and never expected to get.

Trying for college

Unfortunately, that positive achievement couldn't cancel out the big minus of my final year in high school – especially the F grade in trigonometry. After graduation, when my Dad suggested that I try college for six months instead of joining the Army, my school record became a roadblock in my path.

And going to college had become my path. In fact, despite my teenage boredom with high school in general, I had always followed a "college prep" curriculum (including physics and trig). In my grade school days my teachers – and especially my mother – kept telling me how "smart" I was. So maybe, in the back of my mind, I believed it a little. While I guess I grew up wanting to follow in my father's footsteps as an electrician, there was always that "hedge" of going to college in the back of my mind.

However, now that I had decided to go, it was pretty clear that I wasn't going to be admitted to a prestigious, name-brand university like one of the Ivy League schools or a major engineering school (Newark College of Engineering and their ilk). You don't get into those places when you graduate from high school with an overall 78 average.

That was no big deal. Going to any college would be amazing enough. The real problem was that with an F in trig on my record I might not get into any college at all. That was a big deal.

Fairleigh Dickinson University, then a young institution with little recognition outside of New Jersey, came to the rescue. It was willing to let me enroll, provided I passed a summer course in trigonometry. Here I'd barely made it out of high school, and Fairleigh gave me a chance to redeem myself. I'll always be grateful to them.

The FDU advisor who spoke to me about enrollment knew that a college-level trig class had started a week before in the school's evening program. He thought that since I had already been exposed to trig, he might be able to talk the instructor into letting me take the class. The instructor agreed.

Now that I had a real goal, I did what I've been doing ever since. I tried to build on what little I had, and spent the summer studying trig. I wound up with an A, and in Fall 1962 officially enrolled in the electrical engineering degree program at FDU.

I had also started working at an Acme supermarket to earn money for college. There weren't a lot of ways to fund a college education back then. Today's system of student loans and grants just didn't exist. Fortunately tuitions then weren't as onerous as they are now, either. But unless I worked, I couldn't go to college. My parents, like many others at that time, expected that I would work my way through school.

Higher education, higher stakes

I enrolled at Fairleigh Dickinson without knowing what to expect. It was a new phase of my life: no more goofing off, no encounters with the law. At the end of four years I had earned not one bachelor's degree, but two, and was starting on a master's.

It wasn't innate genius on my part that earned me those degrees, believe me. I didn't stun my professors by acing all my courses. In fact, I got nothing but C's my first semester. But by the third semester I was at least on the dean's list.

I trace a lot of this success to my very first day on campus. While standing in line at registration, I met some other incoming students, including several foreign students who were a bit older and more seriously inclined. In their case, doing poorly in class meant a quick trip back to the homeland. So studying was uppermost in their minds. Plus, as with most foreign students, they had a better math background than most of the incoming Americans. It was another grace moment.

I wound up joining a study group that included Moshe Av-Ron, a 26-year-old Israeli soldier, and Jerry Berman, another 17-year-old with a background similar to mine, except that he was raised in the Bronx. We all felt the need to do well our first semester. Later on Bernie Price, 10 years our senior, joined the group. I still keep up a friendship with these guys.

Now, I could have just as easily met fraternity boys on that line and, believe me, I would have been right there with them, partying, drinking, and carrying on. I saw many of these types flunk out in the first semester. It's almost like I have this guardian angel watching over me: when I really need a direction or help in life, it seems to appear in some inexplicable fashion. I don't understand it but, believe me, I sure do appreciate it!

This proves that when your mother says you're known by the company you keep, she's only half right. You hang around with smart colleagues and mentors not just because it makes you look good, but because their example shows you how to get better. It helps if they're the right kind of mentors, not backwoods troublemakers trying to recruit confederates in crime!

Not everything went smoothly, of course. Acme Supermarket fired me for wrecking their delivery car, so I wound up working as a busboy in Sam's tavern for the next four years. Since that didn't bring in enough cash for my tuition, during the next four summers I also worked in New York in my father's electricians' union.

Academic progress

College was an eye-opening experience for me. Growing up in the Poconos and small-town New Jersey made for a pretty isolated environment. At FDU I was introduced to people and cultures from around the world. I learned lessons about the importance and rewards of work that carried over throughout the course of my life.

Originally I'd planned to major in electrical engineering (EE). Then an enterprising professor talked three of my buddies and me into going for degrees in both physics and EE. He worked it so we could earn them within the standard four years.

He wasn't the only positive influence on our lives. Teachers like Lee Gildart, Oswald Haase, Peter Walsh, and Ralph Hautau in physics, and William Schick and Ernest Wantuch in EE, really helped us develop the skills we needed to be scientists. Thanks to them, we came out of FDU ready to compete with anyone, even guys from MIT and Cal Tech. They also reinforced my interest in space.

At the end of the four years, with a BSEE and a BS in Physics in hand, I decided to stay at FDU to earn an MS in Physics.

Graduate school is a lot different from the undergraduate routine. You're completely focused on your major, and you can get involved in professional activities. These include attending conferences and trade shows where experts present papers and you learn about the latest developments in your field.

One of those trade shows, held in 1967 by the IEEE at the old New York Coliseum, altered the course of my life. Frank Palaia, another FDU grad student, went there with me, and we met a professor from the University of Virginia named Avery Catlin. He recruited both of us to go to UVA to study for a PhD. That was a crucial step on the road to my career.

How things had changed! In just six years I'd gone from wondering if any college would have me to being asked to join a graduate program at one of the country's oldest and most prestigious universities. It was beyond gratifying.

It was also another example of grace at work. I'd certainly never thought about going to a school like UVA, and pursuing a PhD was not a goal I'd ever considered. This random stroke of luck led me to a career in materials science.

Things were changing in my personal life too. I got engaged in March of 1968 to Brenda Lechner, and we married that June. Then my new wife and I packed up and went off to UVA so I could study for a

PhD in Materials Science. We lived on my fellowship of $600 a month in the summer and $300 a month during the academic year. Brenda supplemented this income with hospital work.

Virginia was founded in 1819 by Thomas Jefferson, who designed its famous Rotunda building. Partly because of attending UVA I've developed a lifelong interest in and admiration for Jefferson.

It's also a tough, top-flight school. I was lucky to have two outstanding advisors, Professors William Jesser and Doris Wilsdorf, to help me get through.

I studied metallurgy and metal physics, branches of materials science that helped create the electronics revolution. Every electronic gadget in use today gets its basic functionality from the characteristics of the materials it's made from. These were some of the hottest fields of study in the sciences.

And they weren't easy. I did OK in the coursework, but the general (or "comprehensive") exams for the PhD degree were another story. I barely squeaked through. That was another "gift" from life.

By the end I'd been a student long enough. We had a new baby in the house (my daughter Kim, born in August, 1970), and I had to start making a living. So I worked hard and finished my dissertation in January of 1971. I was Doctor Olsen.

Even my Dad was proud of my achievement. Soon after I got my PhD I went to visit him in the Westview Tavern where he hung out in Ridgefield Park, New Jersey. For once he seemed really happy to see me. He went around introducing me to his friends as "my son, the doctor; Dr. Greg Olsen."

Unfortunately, when I started looking for a job, my degree didn't get me very far. By 1971there were so many PhDs in my field that competition for the few positions available was intense. Fate had intervened again, this time handing me a setback.

But Professor William Jesser, my UVA advisor, found a post-doctoral position for me at the Physics Department at the University of Port Elizabeth in South Africa, where I did some teaching and research.

South Africa, of all countries? For a lot of people that would have been like moving to the moon, but I needed a job, so I packed up my family and off I went. In a way, I guess, it was another example of me getting focused on a goal and pursuing it without thinking of anything else.

It turned out to be a great experience. South Africa was a very different place in those days – apartheid was still in force, and relations with the rest of the world, not to mention the rest of Africa, were tense. But it's a unique country, and I grew to like it very much.

So much, in fact, that 30 years later I bought a wine farm there. I just love the land in and around Capetown, the spectacular mountains – I'll put it up against the Napa Valley region of California for scenery any day. South Africa has become a second home, and it's been wonderful to watch its new democracy take root and grow.

I was at the University for 18 months. Finally, in September 1972, RCA Labs in Princeton, New Jersey offered me a job, and my career began in earnest. Hard work and chasing the dream played a major role, but once again grace had intervened, as it had so often in my life. Brenda's cousin Arnold Moore worked at the Labs, and he let her know that after a long hiring freeze they were recruiting again. I applied, and was offered a job by Dave Richmond.

In my view that's the secret to success, however you define it. My case proves it isn't all luck, and it isn't all native talent. Ultimately I got to travel on Soyuz not because I was some kind of superstar, but because I had a dream and I didn't give up, and even when the odds were against me, took advantage of the breaks that fell my way. You can call it stubbornness or determination, but whatever it is, it's a powerful ally.

Most important, my background didn't define me – thank goodness. An average student can still do well by putting in the effort. A rock-throwing, hubcap-stealing kid can do something worthwhile when he sets his mind to it. But like everything else in life, you don't do this alone. You need others to show you how.

Prof. Lee Gildard   *Prof. Oswald Haase*   *Prof. Doris Wilsdorf*   *Prof. William Jesser*

Vladimir Ban and I in 1984, holding the investment check for Epitaxx – nearly $1.5M.

4 A New Vision: Career Launch

Princeton, NJ (August 7-August 20, 1996) — In 1984, while Sarnoff was still the research arm of RCA, Olsen and a colleague wrote a business plan and founded a company called Epitaxx. With $1.5 million in venture capital from Warburg Pincus in New York City and DSV Partners in Princeton, Epitaxx started making fiber optic light detectors. In 1990 Nippon Sheet Glass offered the partners nearly $12 million for the company, and they sold.

— Business News New Jersey

Ten years later the future was looking pretty good. I was re-established as a single guy. My wife and I had married too young, and had grown apart over the years, so we decided to split up in 1979. By today's standards it was a reasonably amicable divorce, and I enjoyed spending as much time as I could with my two daughters.

I was also 10 years into a career in science and technology in the David Sarnoff Research Center of RCA Labs in Princeton, NJ. The Labs was the central research and development (R&D) laboratory for one of the biggest consumer electronics companies in the country. RCA not only made TVs, radios, and records, it was a powerhouse in electronic components, broadcasting (the NBC radio and TV networks), government work, even communications satellites.

The labs were devoted to creating technology that RCA could commercialize. By any standard you want to apply, they were fantastically successful. Color TV, the liquid crystal display (LCD), the semiconductor laser, the process for making integrated circuit chips – these and other basic electronic technologies were invented or commercialized at the Sarnoff center.

Sarnoff was a great place to be if you were a scientist. You rubbed shoulders with some of the leading electronics researchers in the world. The pay was good, and the surroundings were pleasant.

Our research facilities were housed in a labyrinthine brick building with impossibly long corridors, located on a wooded, 300-acre campus straddling Princeton and West Windsor. There was a central cafeteria serving breakfast and lunch, where engineers and scientists from a variety of disciplines discussed projects over meals and coffee breaks.

We did interesting work, though the average person might have considered it kind of nerdy. Remember, this was before celebrities like Bill Gates, Steve Jobs, and people like the founders of Google and eBay had made nerdiness "cool."

My own research involved developing a technique to fabricate semiconductor wafers called vapor-phase epitaxial growth. We were growing indium gallium arsenide (InGaAs) crystals. That may sound like a theoretical pursuit only a scientist could love, but it translates into something important in our daily lives. These crystals are used to create laser diodes and photodetectors, two of the basic building blocks of our wired world.

Today they power worldwide fiber optic communication networks that carry telephone calls, Internet traffic, high-def TV, and everything you can get on the Web, from e-mail and online shopping to FaceBook sites and YouTube videos. Back then, though, most of these services didn't even exist.

Lasers and photodetectors are mirror-image "optoelectronics" gadgets. Apply an electrical current to a laser, and it emits coherent light. Hit a photodetector with light, and it produces an electrical current. They act as send-and-receive pairs in fiber optic networks, using laser light to carry signals over long distances at high speed.

Of course I wasn't the only scientist working to refine InGaAs processes. Colleagues like Jim Tietjen and Ron Enstrom at RCA Labs were prominent in the field, as was Steve Forrest at Bell Labs. But I made some significant contributions.

One of the exciting things about science is that you can contribute to progress just by doing your job well. You don't have to discover a new law of the universe to make a positive impact on the way we live.

And sometimes a career in science can lead you to a bigger opportunity. That's what happened to me. (Knowing people like Steve Forrest is a real advantage when that happens – more on that later.)

Right around the time I was working on these devices, the need for communications networks with higher speed and more capacity was intensifying. I was at the right place at the right time. My research on optoelectronics devices became the basis for founding two highly successful companies, which ultimately gave me the financial resources to book a seat on Soyuz.

First stage: Sarnoff to Epitaxx

Frankly, right up to the point I founded my first company, Epitaxx, I had no intention of being anything other than a scientist. Once again it was pure grace that presented me with an opportunity.

Communicating with light

The opportunity came about because the telecommunications industry needed higher speed and more data and voice capacity, and the only way to get that was to create an entirely new kind of network. Devices like the ones I was working on were destined to play a major role.

Why did the industry need a new network? Because the old phone system was clunky and expensive.

Call London or Beijing? A major hassle, and line charges of at least $10/minute. Access the Internet? Dial-up service, agonizingly slow response, and fees charged for every second you were connected.

We were placing demands on the analog system that it couldn't meet. And the old copper phone lines had reached the limits of their capacity.

The answer was an all-digital network. It would carry voice and data with ease. For maximum speed and capacity it would use fiber optic cable, made of ultra-pure glass, to transmit information on beams of laser light at fantastic speed. This would also immunize the network against electrical interference, eliminating noise and distortion and enabling fast, sure communications.

To make this happen we needed laser-based devices that could convert electrical signals from telephones and computers into ridiculously short flashes of light, like some insanely fast version of Morse code, at the transmitting end. At the receiving end, photodetectors would decode the flashes back into electrical signals.

By 1982 all the pieces of the puzzle were in place. We had fiber optic cable, digital technology, and optoelectronic devices. The sales opportunities were huge. Lots of lasers and light sensors would be needed to switch the worldwide communications system to fiber optics.

The devices I was working on would be a hot commodity. My future was staring me in the face. Yet I was not aware of it. I still saw myself as a research scientist.

Colleagues and mentors

My immersion in optoelectronics research had begun on Sept 11, 1972, when I joined Sarnoff/RCA Labs. Those were the golden years of semiconductor R&D. Every major company – AT&T, RCA, Hewlett Packard, IBM, Rockwell, and dozens more – had its own lab staffed by hundreds of scientists, all working to create the brightest laser or fastest transistor.

I had no practical background in semiconductor electronics, except for some college courses in solid state physics and quantum mechanics. But I'd done my PhD thesis at the University of Virginia on crystal defects in metals. When I started at RCA I learned that the GaAs-like materials they were using to make lasers had the same types of defects that I had studied in my graduate days. So I immediately fit in from a scientific standpoint.

Things went well at the interpersonal level too. I have been told that at some of the larger labs (Bell Labs, for instance) there was intense competition among the scientists, and researchers would not always be forthcoming with help for their colleagues. This was not the case at Sarnoff. Since it had a much smaller staff, the people there tended to work together more and share their results.

At Sarnoff I was lucky to associate with some very smart people who knew a lot about optoelectronics. Sarnoff was home to experts like Mickey Abrahams, Mike Ettenberg, Chuck Neuse, and Vladimir Ban.

I recall meeting Mickey Abrahams my first day. He did electron microscopy, which allows you to see an image of the actual atoms involved in defects found in, for example, semiconductor crystals. This was the same technique I had used at UVA, so we hit it off right away.

And we both smoked, which was another point of compatibility.

Good relationships with colleagues can speed your work. To give you another example, there was the time I needed to take a Polaroid photo of a sample and didn't know where the supplies were located. At a larger lab I might have been left to fend for myself. But I was lucky enough to run into Mike Ettenberg, who had been at Sarnoff for two years by this time.

He had a complete optical microscope setup in his room – including Polaroid instant film, which at that time was an expensive luxury. He could have blown me off and pointed me to the stockroom. Instead, he eagerly showed me his setup, taught me how to use his microscope, even shared his private stash of high-speed Polaroid film. Wow – another lucky day, and another good lesson in the importance of networking at the workplace.

From Mike I learned how to network with others and scrounge for supplies. He knew how to play the system, how to beat it, how to cut corners to reach your objective. And he was generous in sharing all of his knowledge. He also had a quick mind, and I really enjoyed working with him as a bench scientist.

Mike made himself an ally and I did some of my best research with him. We are good friends to this day, but now we share golf tee times instead of Polaroid film.

Chuck Neuse furthered my development by teaching me how to write and make lucid presentations. Chuck was a master at distilling complicated information down to its essence and bringing out whatever meaningful conclusions were there.

He was a maestro of the overhead projector viewgraph, those clear plastic sheets everyone used before LCD projectors and PowerPoint came along, and he was just as good with PowerPoint in later years. Everyone enjoyed his presentations because you always learned something, and you weren't overwhelmed with detail.

Chuck was also a great boss. He worked as team member, never pulled rank, and always treated everyone as an equal. Later on, when I had people working for me, I tried to follow his example.

After I'd been at the Labs about six months, a guy from Croatia named Vladimir Ban returned from a "sabbatical" visit that he had done at Imperial College in London. He had the office right across from mine and did theoretical work on vapor phase epitaxy, the same method I was using to grow my crystals. We hit it off, going for beers after work, sharing our love for scuba diving.

Vladimir went through a divorce shortly after I met him. Five years later he commiserated with me when I went through mine. Mentors come in all types, at all times of your life. The shared experience of divorce was probably one of the factors in our joint decision to start a company together in 1983.

Speaking of mentors, at Sarnoff I met somebody who is still a real "go-to" guy for me whenever I have serious issues to resolve. Henry Kressel is a no-nonsense person who doesn't mince words. You may not like what he says, but you can be sure that you will get the plain, unvarnished truth as he sees it. He was on the advisory board of both of the companies I founded.

Henry built an awesome record as a scientist. When he started his research on lasers, they were laboratory curiosities. Most of them burned out after a few seconds of operation. He led the development of the first commercially viable devices, with lifetimes measured in tens of thousand of hours. In a very real sense we owe many of the benefits we derive from laser technology to his work.

Henry favors an abrupt style of communication. In 1976, Vladimir Ban, Mike Ettenberg and I drove up to a conference in Ithaca, NY. In those days it was not uncommon for a bunch of guys in a car to carry a couple of six-packs, and consume them during the trip.

When we reached Ithaca, we decided to visit Cayuga Falls, one of the local scenic sites. On the way we ran into Henry, who had just arrived at the conference, and invited him along. I remember his having to step over 12 empty beer cans while climbing into the back seat, and I vividly recall the evil-eyed look he gave Mike, who was his direct subordinate (his "serf," as we used to call him).

It was a 20-minute ride to the falls. Henry got out of the car and looked around for about 30 seconds. "Nice view….really nice view," he said. Then he got back in the car and said "Come on, let's go." We all dutifully rode back to the conference.

Henry was regarded as quite a taskmaster. But his competence and sheer quick mind were unsurpassed and we all respected him immensely.

Patents and portents

After a few years at the labs I managed to develop some name recognition of my own in the optoelectronics field. As was expected of a research scientist, I wrote technical papers and gave talks on the work I was doing. I did over 100 of these papers during my career.

I also earned 12 patents in the electronics field. Many of these are modifications of existing ideas – nothing that could be considered Nobel Prize material. Any honest researcher with multiple patents would tell you the same.

But every researcher has at least a couple of patents that we think are really good ideas. In my case, it was the so-called "double-barrel vapor

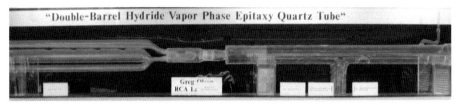

In 1978 the US Patent Office rejected the patent application for my crystal growth chamber. Six years later that action saved Epitaxx. The original chamber still hangs in my office.

phase epitaxy tube" (see photo). I'm not even going to begin explaining it, except to say that it was a better way to grow certain types of electronic devices, like lasers and photodetectors, the kind of parts RCA was trying to sell in the early 80's.

I had designed the tube with my assistant Tom Zamerowski. We decided to use two parallel tubes instead of one to speed up growth while decreasing the number of defects in the lasers. It was called a "double-barrel" tube because of its resemblance to a double-barreled shotgun.

It was really fun, drawing this thing up, having it built in the various shops at RCA Labs, and actually seeing it working. My creative juices were flowing in those days.

I still think that tube was a good idea; it certainly was in 1978. But the patent office didn't agree. I submitted a patent application through the labs, and even after appeal, the patent examiners turned it down on a technicality. It didn't have much to do with my tube, but ... well ... try changing the government's mind once it's made up.

I was bummed out and depressed. Here was one of my best ideas and the patent office says "... sorry ... rejected." Well, another life lesson learned.

This disappointment turned out to be one of best things that ever happened to me. If the patent office had granted RCA that patent, I probably never could have started my first company, Epitaxx, as we'll see a little later.

Making it work

Patented or not, the tube proved invaluable in my research. But there was a problem getting it installed.

I think it was in early 1979 that Tom and I finally set it up, only to find that it was a little too big for the room where we planned to use it. We needed about an extra 12" of space, so that we could move a half-inch-diameter quartz rod in and out of the system.

The room had a window, so I figured, "Hey…let's just cut a hole in the glass pane so the tube can stick out the window. That will solve the problem." That's when the bureaucracy kicked in. (Bureaucracy exists everywhere, even at R&D labs.)

The facilities manager quickly pointed out that putting a hole in the window glass violated several building codes and could potentially upset the air balance in the room. But once again I persisted and followed my mantra: don't give up!

It nearly took an act of Congress to get a simple 1″ diameter hole drilled in the glass window, with a secure shutter to keep it closed at all times except when growing crystals. And you know, after fighting me on it for weeks, several of the facilities people came around to admire – and joke about – the "phallic" protrusion from the vapor phase epitaxy window!

It should also be noted that we were working with a very toxic and deadly gas called "arsine," which contains the arsenic used in gallium arsenide crystals. Just a small whiff of this gas can be fatal. We all knew what we were dealing with, and I must say that there was never a serious accident during my tenure at RCA Labs – nor at any other time to my knowledge.

Nevertheless, I remember that some of the early VPE systems, circa 1972, used quartz tubes to carry the arsine! One slip of the elbow, and a serious accident could have resulted in spite of the ventilation hoods. We simply accepted the risk, just as astronauts accept the risks of flying into space.

In today's world things are much better from a safety point of view. Arsine gas is treated with a lot more respect, using double-walled steel pipes, parts-per-billion gas monitors, and careful leak checking.

Pioneers always face hazards that are eliminated by the time the general public follows the paths they have blazed. But the risks are worth it. Anyone who uses a cell phone carries one of these GaAs chips, which most likely started as arsine gas.

Resources for the future

I'm proud of my achievements as a researcher. At the same time I know that while I was pretty good as an R&D scientist, I was not the best.

However, by giving papers and attending conferences I got the chance to meet some of the best. I was hobnobbing with researchers from Bell Labs, Rockwell Scientific, and other leaders in the field. One of the contacts I made at these events, Steven Forrest from Bell Labs, was a brilliant physicist whose innovations made the manufacture of In-GaAs detectors practical. He later proved to be an important resource for Epitaxx and for my second company, Sensors Unlimited.

The competition among the companies was fun, and the camaraderie at the conferences where we shared our results is something I miss to this day.

All of this work gave me the technical knowhow to make an opto-electronic product, and the industry background to understand the market. But I still wasn't thinking about striking out on my own. I wasn't even thinking about trading my white lab coat for the dark suit of a Sarnoff manager. I've never been an organization man, and wouldn't have been much good at climbing corporate ladders.

But in 1982 I got the first gentle push toward the exit. Sarnoff's parent company, RCA, decided to turn my InGaAs photodetector into a real product. As a prime developer, I was responsible for transferring the technology to a manufacturing plant in Montreal.

Because RCA was used to doing everything on a large scale, they made the process of producing the devices very complicated. Watching this unfold, I was sure that I could do it "faster, better, cheaper."

To RCA the photodetector was no big deal – just a $1 to $2 million market. Not that there was anything wrong with that, but as a major corporation they were not going to pay a lot of attention to new, developing markets. To me, however, a $2 million market didn't seem small at all.

The Montreal experience was an eye-opener. It showed me that a big company isn't necessarily better at implementing an idea than a small one, especially in a new technology. The idea of leaving a cozy lab job to start a new company was suddenly a lot less scary.

There was encouragement from a couple of other directions, too. For example, Vladimir Ban gave me a book on how to start your own company. Even though I wasn't thinking along those lines yet, I read it on his recommendation.

Pitching for dollars

Then there was Andy, an acquaintance who worked at W.R. Grace, the New York-based global chemicals and materials company. I met him because I was dating his sister-in-law, Laura.

The four of us, Andy and I and the two sisters, used to go out for dinner. Andy, who worked in the venture capital division of Grace, would push me to "think big" and start a company. He told me Grace was giving money to anyone with a decent idea for a high-tech business, and I should apply. He was convinced I must have some brilliant plans that would impress the venture capitalists.

I don't know about brilliant, but by that time I did have an idea. I knew the telecommunications industry pretty well. And I was pretty sure an optoelectronics provider could ride to success on the coattails of the coming digital transformation. Starting a company was beginning to sound like a good idea.

But frankly I had no clue how to do it. As a career researcher I had no experience with the challenges of entrepreneurship or the complexities of running a real business. So I did everything by the seat of my pants. This has its pitfalls, as I discovered when I went looking for money to finance the new company.

Early in 1983 I went through three rounds of fund-raising. Every

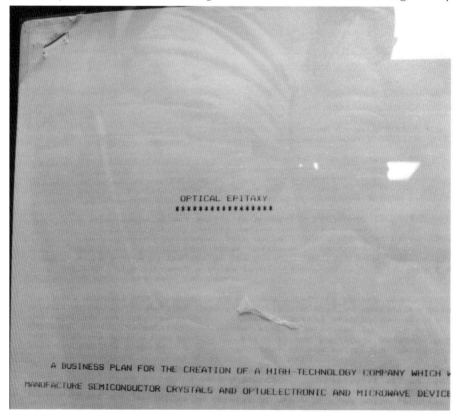

OPTICAL EPITAXY
****

A BUSINESS PLAN FOR THE CREATION OF A HIGH-TECHNOLOGY COMPANY WHICH W
MANUFACTURE SEMICONDUCTOR CRYSTALS AND OPTOELECTRONIC AND MICROWAVE DEVICE

Original business plan for Epitaxx, typed by my 13-year-old daughter Kimberly.

round was an educational experience, especially the first two. (That's another way of saying "qualified disaster.")

The initial pitch for funding happened when Andy got me an interview with the W.R. Grace venture capital team. I explained that my idea was to make photodiodes for optical fiber communications and cable TV transmission.

Of course the first thing they wanted to see was a business plan. That's the basic starting point in creating a new company. Every would-be entrepreneur has one.

Except me. I said, "A business plan? What's that?" The conversation was pretty much over right then.

It was a humbling experience, but I didn't give up. Over the next few weeks, working at home after hours, I wrote a business plan for my new company.

My 13-year-old daughter Kim typed the final version for me in July, 1983 using a borrowed HP word processor. It took her maybe 20 hours. Poor Kim: she got to spend an entire weekend deciphering her father's scrawled handwriting and entering it into the word processor, while I took her younger sister Krista for a visit to the Cloisters museum in Manhattan!

This was when personal computers were practically unheard of, and most people were creating documents on typewriters. Using a dedicated word processor definitely qualified my daughter as a high-tech guru. I still have the original copy of that plan, framed alongside news clippings about the $12 million sale of the company in 1990, hanging on my office wall.

When the plan was done, I asked my buddy Vladimir Ban to review it. He liked it so much he offered to join the new company. We decided to be equal partners, with Vladimir as chief technical officer (CTO) and me as chief executive officer (CEO).

This was a real vote of confidence on his part, because leaving Sarnoff and RCA was going to be harder for Vladimir than for me. He had moved up into the management ranks and had great potential for career advancement.

I had to make some salary guarantees to reassure him he was making the right choice, but there's no doubt he was a risk-taker. He'd earned his B.S. in chemical engineering from the University of Zagreb in Yugoslavia, then come to America to do a PhD in solid state science at Penn State. That's the kind of leap few people attempt.

It was at this time that I got an object lesson in keeping your plans private until you were ready to act on them. We made copies of the business plan, and Vladimir inadvertently left the original in the photocopier – at the labs! One of my colleagues found it and returned it to his desk with a note saying, "You really shouldn't leave this kind of thing around." Point taken….once again, disaster averted!

Business plan in hand, I was ready for the second round of financing talks. My first stop was at Judson Infrared, a small manufacturing company in Phoenixville, PA. I tried to talk them into supporting my concept for the new company, Epitaxx.

I figured that Judson would be receptive to my ideas, because they were trying to do something similar, using older technology. I also knew they had the equipment I needed to manufacture the devices.

To me it seemed logical to create a joint venture built around their equipment and my technology. But during our talks it became obvious that they were more interested in hiring me for their project than in partnering to build a separate company. It was another dead end.

Over the years a number of companies (and people) have turned down opportunities to work with me. There's no way of knowing how things might have turned out – for me or for them – if we'd joined forces.

Getting funded

Finally, in the summer of 1983, Vladimir and I got an audience with a big venture capital firm. It came about through the good offices of Henry Kressel.

Henry had left the labs earlier that same year to join Warburg Pincus in New York. Since he was such a great mentor, I had made an effort to keep in touch with him. I got him to review the Epitaxx business plan.

By that time the plan was a lot more polished. It was 15 pages long, single spaced, and much better thought out. Our strategy, as outlined in the plan, was to focus on a single product (photodetectors), become the technological leaders in that product, and make it so cheaply that even major producers of telecommunications systems wouldn't balk at buying it from a relatively unknown company.

We deliberately chose not to make lasers. They were complex devices, and Epitaxx wouldn't have the resources to design and manufacture them. In addition, several major companies had invested a lot of money in developing these glamour products. If we tried to compete with these behemoths, we'd be crushed.

Because of his technical expertise, Henry understood our concept right away. He got us a hearing with the funding committee.

When we met the team, one of the VCs asked, "What's the elasticity of the market?" I felt like I was back at W.R. Grace – I didn't know what he was talking about, and said so. He explained that it was financial-speak for "what would happen if you reduced prices to increase your sales?"

Fortunately that misstep didn't kill our bid. Perhaps it was because our funding request must have seemed like pocket change to a VC firm. Warburg was used to investing tens of millions of dollars in a company. We were asking for $1.7 million.

Then we went back to await their decision. I had taken the afternoon off and was relaxing. I remember being so calm, half asleep, just lying across my bed, when the phone rang. Normally I wouldn't have thought of answering it, but this time something told me to pick it up.

It was Henry. He said: "Listen, we decided to fund your plan….be at my office at 9AM tomorrow….." and hung up.

Anyone who deals with Henry is familiar with these no-nonsense 12-second phone calls. He gets his message across – no need for small talk.

At the meeting they offered us $1.5 million in exchange for ownership of half of the company. Our lawyer thought those terms were pretty hard-nosed: Warburg would get 50% of Epitaxx for very little money.

I had a different outlook. Nobody had ever offered me $1.5 million for an idea before, and if they wanted to give me that much money, I was going to take it.

When the check was ready, Vladimir and I returned to New York to pick it up. It was typed like an ordinary business check.

To celebrate, we went to a Manhattan bar for a beer, and kept passing the check back and forth between us. This was more money than either of us had ever seen, and somebody had hand-signed a check for it, the way you'd sign a check at the grocery store.

I even remember showing it to one of my neighbors as I was leaving for work the next morning. I saw her coming out of her house and shouted, "Hey, have you ever seen a million dollars?"

"Get out of here," she said. "You don't have a million dollars!" Her eyes nearly popped out of her head when I showed her the now wrinkled check for $1.47 million.

From today's perspective our funding was pretty paltry. Just compare what we got with what VCs were handing out to dubious dot-com companies – or how much I was able to sell my second company for – at the turn of the century.

But it was enough to get us started, and that's all that counted. Besides, it's not what resources you have, it's how you use them. I was committed to building my own company, using all the technical skills I'd worked so hard to acquire. I was sold on the idea of being an entrepreneur, and threw myself into it.

High tech on a low-tech budget

In November of 1983 Vladimir and I left Sarnoff and the big RCA "family" to set up our new company. The amazing thing is, although a number of people there warned me that I was leaving a secure job and taking a big risk, *they* were really the ones taking the risk.

They were counting on things staying as they were. But just three years later GE bought RCA, split up the company, and laid off a number of lab personnel. You could argue that I was actually making the safe move by leaving when I did.

I'd like to say that the rest is history, but nothing is that easy except in storybooks. To make the most of an opportunity, you have to keep working at it.

Our first challenge was to find a place for the business that included office and manufacturing space. Irwin Kudman, an ex-RCA guy and mentor, gave us a small office in his hi-tech business while we looked for a site. Irwin was one of the first of the many entrepreneurs that came out of RCA's Sarnoff Labs.

It wasn't easy. We needed clean room facilities to grow the crystals and make the detectors. The slightest speck of dust or other contaminant in the air would ruin the product. But we had to keep the cost under a million dollars. That's a tenth of today's going rate – or less!

We found the unlikely solution when one of Vladimir's family friends, who was active in Princeton real estate, introduced us to a local legend: Mr. Ted Potts. He was an octogenarian businessman who owned a lot of property on Route One. His biggest holdings were a couple of older business parks with modest but flexible office spaces.

Mr. Potts had a dry sense of humor and some unusual business methods. Faced with a bill he didn't think was fair, he said he'd "just slip it in the bottom drawer and leave it there." Once I asked him for advice on whether to get out of a contract I'd signed to buy a condo that I really didn't want. "Don't let the sun set one more day on that contract," he counseled.

When asked by a salesman for a line of venetian blinds why he insisted on staying with his present supplier, rather than buying the salesman's "clearly superior product," Mr. Potts replied, "because I know how bad that guy is … I can't say the same about you!" He was a great business mentor for a 38-year-old scientist with little real-world experience.

He was definitely old-school (he was Mr. Potts to everyone who knew him, and he always called me Dr. Olsen), but he was open to new ideas. He found the idea of a high-tech manufacturer renting space alongside such usual business tenants as advertising agencies, financial counselors, and accountants exciting. And nothing fazed him.

Did we need to locate tanks of hydrogen gas outside our building, and pipe the hydrogen through the walls to our equipment? Shades of the hole in the lab window that prompted my battle with Sarnoff's facilities crew! Other landlords would have walked away, but for him it was no problem.

He also took disasters in stride. When I got a call at 7 a.m. on New Year's Day 1990 that a water pipe had burst in our facility, I raced there to deal with the mess, only to find Mr. Potts already on site, directing a clean-up crew.

Space at his business park was affordable, but about as low-tech as you can get. To adapt it to our purposes we had to improvise a low-budget clean room. We were going to find a way, no matter how unconventional, to get the job done.

Instead of costly dust-free panels for the clean room walls, for example, we used regular sheetrock that we coated with epoxy paint. We couldn't afford clean-room-rated flooring, so we found some standard floor covering that was good enough. Mr. Potts provided referrals to suppliers and contractors who could help.

Overcoming adversity

Finally, after months of cutting corners and hard work, we were ready to make our detectors. We officially launched Epitaxx in April 1984 with five employees. We promoted our products through the normal industrial channels, such as advertising in technical journals and exhibiting at trade shows, and we started to make sales calls to potential customers.

Our first sale was to the University of California – Berkeley, at $600 per detector. Today those detectors go for $1 each.

It was just around then that the first bump appeared in the road, and it came in the form of a letter from RCA's legal department.

When I had left the previous November, I got the typical sendoff. Most people wished me well, my buddies and friends threw a luncheon for me, and I was even invited back for our group's Christmas party a month later. No one at the Labs seemed to have any objections to my starting a small optoelectronics company.

No one, that is, until two other guys decided to leave in March, 1984 to join yet other startup companies in the optoelectronics arena. Now RCA felt that the floodgates were opening: all their star engineers, whom they had trained and in whom they had invested millions, were running off and starting companies. Their lawyers got busy and letters went out.

The next few months were very stressful. Any entrepreneur who has opened a registered letter from a law firm beginning with the words "You MAY be in violation of your employment agreement" knows exactly what I am talking about. It is a scary prospect to be on your own, without significant resources, up against a billion-dollar company with a staff of 50 full-time patent attorneys.

Whether or not you are actually in violation is probably irrelevant. The mere fact of receiving such letters has, I'm sure, discouraged more than one potential entrepreneur from going on. But me? I've lived with stress most of my life, and while I will admit the letter did have its intended chilling effect, I went into my usual hedgehog mentality. I thought, "OK ... another troubling event in my life ... I'd better put my head down and get through it...."

Even though I didn't think we were violating either RCA patents or my employment agreement, reasonable men often differ on such points, and that is what keeps lawyers employed. A series of meetings followed, including some with lawyers. A number of people who had once been co-workers, even friends, were now looking with angry faces across the table at me.

RCA's legal team for this action included William Burke, the Sarnoff labs patent attorney. We'd known each other at the labs, and had gone out for a beer or two after work. Bill has a PhD in physics to go with his law degree, but he's a down-to-earth guy with a great sense of humor that's enhanced by his pronounced Boston accent.

Sitting across the table from me, however, he wasn't his usual good-natured self. In fact, his attitude was tough and threatening. We might have been friends, but he had a job to do.

After a time, my gut started telling me that RCA wasn't necessarily out to shut us down. They wanted to feel that they were asserting their rights, that they were putting me in my place, and that they were setting an example for other possible entrepreneurs who might be thinking of doing what I had done.

I came to the realization that if we conceded some of their demands, we could still have our company, mostly as we had envisioned it. It became clear that their major concern was our use of the double-barrel vapor phase epitaxy tube for growing crystals that I had developed in 1978. I began to think that if I gave these guys their "pound of flesh" – if they could walk away with some type of agreement that justified their cause – we might not be in such a bad position.

It is a lesson I have learned in life time and time again. Sometimes you've got to lose to win. You have to give up a little, or pull back from your own demands, in order to get where you want to be.

So we reached a settlement whereby Epitaxx agreed not to use certain embodiments of the crystal growth tube. This hurt, as we had already incorporated them into our system. But being experts in our field, it was easy for us to get around this restriction without violating the agreement.

In a win for our side, the agreement further stipulated an embodiment of the tube that we could use, but RCA couldn't!! So RCA got us to knuckle under, we signed their "punitive" agreement, and life went on. In two months we were making crystals that were better than we'd ever managed on our old machine – the RCA-based machine, that is! The final solution was almost comical, but it worked for both sides and I must say that, all in all, the RCA people acted like the gentlemen I knew them to be.

This "tube" story had a funny ending. When we signed the agreement, RCA insisted that we destroy the two existing tubes that we had made using their design. Furthermore, they wanted to witness the destruction of each tube.

So at the "signing" ceremony, we placed one of the tubes in a box and broke it in several places. But they didn't think that was good enough. So we called in one of our techs, Mike Popov, a guy with a burly physique like you used to see on Soviet propaganda posters. He grabbed a ball peen hammer and proceeded to pound the tube into smithereens. When this five-foot glass tube was essentially reduced to powder, the attorneys walked away, smiles on their faces. Were they amused, or simply satisfied?

If there's one thing I learned from this episode, it's that what seems like a setback can turn out to be for the best. I was really disappointed back in 1978 when I didn't get a patent on the tube. But that rejection may have saved my company. When push came to shove, RCA really didn't have any intellectual property grounds to stand on when our conflict arose. No wonder they settled as they did!

There's another lesson, now that I think of it: work is the best way to keep yourself sane when things are not going well. When the first letter from the RCA lawyers arrived I could have reacted by letting everything grind to a halt while I concentrated on the threat. But my ability to block out problems and focus on what I was really interested in – my work – paid dividends. It let me move ahead according to plan in spite of the situation.

Talent will tell

Back at Epitaxx sales began to ramp up, and in 1985 we actually made a profit. I brought on more technical talent, including Yves Dzialowski, who had earned his PhD in optics from the University of Paris and had worked at some of the early pioneers in optical fiber, including General Optronics and Photodyne.

Yves took on the role of quality control enforcer, a tremendously important function in a young company like Epitaxx. He was always hectoring us poor researchers to pay more attention to quality assurance in the manufacturing process. I remember him saying in moments of very French frustration, "You propellair 'eads wiiill nevair understand ze qualité!"

I also tried to hire Steven Forrest, my acquaintance from Bell Labs and a brilliant researcher and innovator in InGaAs technology. In fact, Steve had developed the process that made it practical to manufacture the detectors I was selling. But he decided instead to go to USC where he continued his research into optoelectronics and organic semiconductors. He did agree to consult for us – fortunately, as it turned out.

When I go looking for new talent, I always tried to find the smartest, best people out there. If they're smarter than me – and there are a lot of people who are – so much the better. It isn't about ego, it's about getting the best people for the job. Yves and Steve both made huge contributions to the success of Epitaxx.

We were doing a lot of product development, and Vladimir Ban was heavily involved. To get one new product, a planar detector, ready for production, he drove back and forth between Epitaxx and a processing partner two or three times a day for weeks. It was an hour each way. To this day he says, "Only small companies can be that pragmatic and that tenacious, and that's what we were."

In 1986 we slipped back a little, barely breaking even. Still, my dream of building a successful company seemed to be on track. But disaster struck in 1987. Customers were telling us that our detectors would work fine for a while, then fail as moisture from the air got into them. The customers' systems were supposed to be 99.9% or more reliable, and our detectors were shutting them down. Epitaxx sales plummeted.

We couldn't find a solution no matter what we tried. But instead of conceding defeat, we treated adversity as a challenge, and turned to our consultant, Steve Forrest, for help.

Steve worked with us by phone, express package, and weekend visit from California, often spending nights on a couch at my house. He analyzed the problem and suggested a whole new way of making the detectors to shield their components from air and moisture. And he helped us implement the solution in production.

Having the best talent in the world on your side certainly paid off. Steve's idea worked. We began shipping the new version, and the failure problem disappeared. Epitaxx went from near failure to being a reputable supplier of reliable detectors in about eight months.

Selling a dream

By 1989 Epitaxx had 55 employees and $5 million in revenues, and it was profitable again. It had grown rapidly and expanded into long wavelength detectors and receivers used by companies to build optical fiber communications networks.

Inside the company things were not so good, at least not between Vladimir Ban and me. Vladimir and I had been good "drinking buddies" since 1973 – we nursed each other through our divorces, went scuba diving together, and more. As co-founders, however, we didn't work together well at all.

I was supposed to be the outside guy while he was inside, but by the end we were hardly talking to one another. That wasn't good for us or the company. Why this happened is hard to say. Maybe our backgrounds were too similar.

Anyway, our relationship became quite strained during the Epitaxx years. Once we stopped working together we drifted back to the old relationship and today we're good friends again. But in 1989 we just weren't clicking.

One year later, in 1990, we decided to sell Epitaxx to Nippon Sheet Glass America (NSG), a Japanese company, for $12 million. Remember, we had started with barely $1.5 million. This was a triumph. We had increased our value nearly tenfold in just six years!

As I recall, we met NSG through a "new" investment banking firm started by Sandy Weill (Sanford Weill Associates). Our investors had picked them out to see if there might be a match for us.

The thing that surprised me about our negotiations was how fast the Japanese worked – like lightning! They sent in teams of accountants for due diligence and seemed in a real hurry. What I didn't know at the time was that they had just been involved in a long acquisition deal with another, larger telecommunications company that fell through at the last minute. The orders from Tokyo were that this deal had better not fail! It's that grace thing again….

This was the era when the Japanese seemed to be buying everything. Japanese companies had bought Pebble Beach and Rockefeller Center. Everybody assumed that Japan was going to rule the world economically – and they did for a few years. (Think India and China circa 2008 … I wonder if the same thing might happen to those countries in a couple of years?)

As a result of the sale I had enough money that I didn't have to think about working for a while. But there were still some things I wanted to do. And I didn't know it then, but I'd need more resources if I was going to make my space trip.

Starting Sensors Unlimited

After NSG acquired Epitaxx, they implemented a "consensus" management environment in which all managers were considered equal and decisions were supposed to be made by general agreement. It's a very Japanese idea, and works beautifully in the polite but hierarchical culture of Japan. Maybe it could work here, too.

But it didn't fly at all at Epitaxx. Relations between the other managers and me had already become strained, and having to work by consensus made things worse. Once the seeds of dissent are sown within an organization, things go downhill pretty rapidly. People choose sides in disputes, and sometimes use a negative situation for personal gain. I have to admit, I didn't maintain the greatest communication with my management team and I'm sure I made mistakes of my own.

I am grateful to have avoided this sort of thing at my next company, Sensors Unlimited. While we did have disagreements from time to time, it was never personal and we all remained good friends.

By the fall of 1990, NSG saw that the consensus model wasn't working, and proposed to me that I go back to being CEO and president.

I knew that this would require a long-term commitment on my part, plus it would further exacerbate my relationship with the other guys. So I decided to adopt an "exit" strategy. I would stay as a consultant for about a year and then leave. NSG would then bring in a Japanese president (which they probably should have done in the first place).

I must say that the NSG executives were totally honorable and always tried to work with us. They never imposed solutions on the company. This civilized attitude extended to my exit contract. It allowed me to write government funding proposals for myself, as long as they were non-competitive. It even gave me the right to produce imaging devices of my own. Thus the seeds for Sensors Unlimited were sown!

I sort of hung around Epitaxx for the next year. The most useful thing I did was to write and submit six Small Business Innovation Research (SBIR) proposals to the Department of Defense. These were intended to get some R&D money to fund a new company through which I could pursue my latest interest, InGaAs array detectors.

Nothing ever stopped me from being curious, and array detectors piqued my curiosity. In some ways it was like starting Epitaxx all over again. The new company would produce array detectors for larger firms, which would incorporate them into industrial systems for astronomy, spectroscopic analysis, and other applications. The devices would be either one or two-dimensional arrays of individual InGaAs detector elements (like the ones Epitaxx produced).

The new company was technically incorporated in October 1991. I really wasn't seriously thinking about starting a company yet. At this point it was just a name on a piece of paper Its only funding was a $1 million line of credit that I had to personally guarantee with my Epitaxx payout.

My consulting contract with Epitaxx ended on January 31, 1992. I left, not knowing what the future had in store for me.

In principle, I had what most people dream about all their lives: financial independence, kids in college, no real responsibilities, no need to work. My buddy Lubek Jastrzebski loaned me his condo in Clearwater Beach, FL for a week while he was away.

Life should have been perfect. But there I was, on a warm beach in February, watching the sun set, and worrying about what I was going to do with my life! I felt really insecure with nothing to occupy me.

Another career transition

Back in New Jersey I did some half-hearted consulting in February and March, and also interviewed with Sarnoff to be CEO of one of their spinoff companies (I think it was the one called Sensar).

Sarnoff's CEO, Jim Carnes, and executive VP Curt Carlson were very persuasive guys. By the day of my third interview I was seriously considering the position. After all, there was nothing else on the horizon.

I have a vivid recollection of that day in March. I left the house and rode down my driveway, stopping to pick up the mail on the way. There was a bunch of letters in the mailbox, including three thin envelopes from the Department of Defense.

I immediately assumed that these were the rejection letters for my SBIR proposals. I was familiar with DoD notifications from my Epitaxx days, and knew that an acceptance letter was thick, since it included many forms to be filled out. A rejection was one sheet of paper.

Dejected but resigned to my fate, I threw the mail on the passenger seat and proceeded on to the Sarnoff interview, which would probably end with me accepting their offer. That's when fate, or grace, or just dumb luck stepped into my life again and changed all my plans.

At the intersection of Route 1, just a couple of blocks from Sarnoff, there was a particularly long red light. More out of boredom than anything else, I reached over and opened one of the DoD letters. It read: "…we are pleased to inform you…all forms will be sent at a later date…."

Excited but still skeptical, I reached over and opened the other two letters. Sure enough –two more notices of funding awards!! Now I knew what the future was.

Sensors Unlimited was cemented in place at that very intersection. I hope that I concealed my delight while I was turning down a very generous offer from Sarnoff.

Then I headed over to see if my former landlord and mentor Mr. Potts had a place for me to put my new company. It just so happened that Epitaxx was moving to new quarters, and he was looking for a tenant to fill their space in his Route 1 complex. As I remember, he offered me my old digs back on a five-year lease, with free rent for the first year. Of course I took it.

As it turned out, I eventually won all six of the SBIR programs that I had applied for. This was an amazing result, even in 1992, when the funding system was not quite as competitive as it is now. The nearly half-million dollars in "free" SBIR venture capital, plus my Epitaxx payout, was the only startup money ever taken in at Sensors Unlimited.

Country dancer

You can't work all the time, and over the years I've found several leisure activities that I enjoy. Or maybe they found me. That was how, around this same time, I discovered one of the big passions in my life – country and western dancing. That might seem like a strange hobby for a Jersey kid, but it's true.

As I was preparing to leave Epitaxx I got a call from an old crystal growing buddy, Bill Bonner, who had just gotten the "golden handshake" from Bell Labs. They had closed his division, one of many that they eventually shuttered.

Bill was angry and down, so to cheer him up I offered to take him to a topless dancing bar in Manville, NJ called "Frank's Chicken Farm."

Neither one of us knew where it was, but I used to drive past a C&W place in Manville called the Yellow Rose on the way to pick up my kids. I told him to meet me at the Yellow Rose and that we would get directions from the bartender there.

I arrived first and ordered a beer. He joined 10 minutes later, and as I was asking directions from the bartender, they began a country dance lesson. Two women who needed partners literally dragged us off our stools so that they could take the lesson.

I instantly fell in love with country dancing and continue to enjoy it. To this day I have never been to Frank's!! Chalk this up as one of the many things I've done in life that were not the result of long-term planning, but of some sudden, unsuspected encounter. Like my space flight….

Key people involved in Finisar Corporation's acquisition of Sensors Unlimited in 2000.
Back Row: Richard Capalbo, John Soden, Maryfrances Galligan, Jerry Rawls, Gregory Olsen
Front Row: Dennis Sullivan, Judy David, Nezahat Gultekin, Holly Dansbury

5 Sensors Unlimited: Life in Orbit

Sensors Unlimited sold in $600M deal
West Windsor – Sensors Unlimited, a 9-year-old home-grown company that developed an innovative fiber-optic technology, is being sold in a deal valued at more than $600 million…The company was named one of the 50 fastest-growing technology companies in New Jersey by accounting firm Deloitte & Touche last year.

— Mary Sisson, Trenton Times, 8/17/2000

Sensors Unlimited started as an idea and grew into an industry leader. At the end of its first eight years it attracted a buyer who paid $600 million for it. Getting there took a lot of hard, grind-it-out work and some strokes of luck, the tried and true formula for success, at least in my experience. There were also a couple of serious setbacks, and we just barely managed to get through the window of opportunity before it closed. We never gave up. Everyone on the team was committed to making it a success.

You could say that the whole Sensors adventure was like a trial run for a space flight: lots of grueling preparation followed by a fantastic launch that made it all worthwhile. So I guess it's fitting that the success of Sensors gave me the means to go into space. It was an adventure for all of us … and it's quite a story.

Building the launch vehicle

My original idea for Sensors was to make array detectors using InGaAs devices (like the ones Epitaxx produced) for imaging purposes, like cameras. You can arrange a number of individual detectors in a straight line to build a one-dimensional array, or line them up in rows for a two-dimensional version. I wanted to do 2D arrays. Light detected by each sensor would be one pixel of a complete image.

Digital cameras use the same basic principle. When you take a snap-shot with a digital camera (it's hard to believe that just 10 years ago everyone used film), the image is captured by millions of sensors. The difference is that all of those sensors are elements of a single chip, while we were going to use individual sensors lined up in an array.

Also, our sensors detected infrared light. They were targeted at in-dustrial and scientific use. Your vacation pictures would look pretty eerie if you took them with a Sensors array. My initial plan was to supply the arrays to companies that would build cameras around them and sell them to the ultimate users.

I couldn't do this alone, of course. I knew how to make the sensors and arrays, but not a camera, which we might do in the long term. The company needed a technical director, and it needed a processing guru who could help us ramp up production.

So I did what I always do: find the smartest, BEST people available to fill these roles. Having the right people in a startup company is just as important as having the right product – maybe even more important. In fact, people are the key to a successful company, in my opinion.

Our first priority had to be the research programs funded by those six SBIR grants. To fill the technical director slot, I turned to Marshall Cohen, a bright guy who was "between jobs," as they say, and asked him to manage the programs on a consulting basis.

Hiring a co-founder

Marshall was an industry veteran. I'd first met him around 1980, when he was working at Rockwell and I was at Sarnoff. We reconnected in 1988, when he moved to EG&G Princeton Applied Research (PAR). His group at PAR was developing an optical spectrometer, an instru-ment for displaying electrical signals, that used Epitaxx photodiodes. In fact, Marshall was one of the first Epitaxx customers.

Over the next three years Epitaxx and PAR worked together on a number of projects. One of them was an InGaAs linear focal plane array (FPA) that PAR hoped to use in a detection system for near-infrared spectroscopy. Astronomers use this technique to study the atmosphere of distant stars. There were other applications too.

In 1991, during my interregnum period with Epitaxx, Marshall in-vited me to lunch at a Denny's on Route 1 near Princeton and proposed to start a business to do industrial spectroscopy based on these arrays.

He even gave me an initial business plan, which I kept and gave him as a present when he left Sensors in 2006.

Nothing much came of that particular meeting (except some minor indigestion), but two minds were thinking alike. Reading his plan helped spur my thinking about what I was going to do after Epitaxx.

In 1992, after winning all those SBIR programs, and just as I was getting Sensors Unlimited under way, EG&G shut down PAR. Marshall was out of a job. Closing down a division is sometimes unavoidable, but I couldn't believe that EG&G was letting a talent like Marshall go too.

Their mistake was my good fortune, and I took the opportunity to bring him on board. That is how I define luck – taking advantage of opportunity. I think we talked for only about five minutes to figure out a relationship. He agreed to start as a consultant, and to come on board as an employee as soon as we could afford him.

Although Marshall actually joined the Sensors staff in early 1993, after I'd officially started the company, I consider him a *true* co-founder. By sharing his business plan in 1991 he had helped me crystallize my thoughts about the direction I wanted to take. We both wanted to work with this technology, and we were determined to make it go.

I quickly realized that Marshall was smarter than I was and in many areas more effective. Instead of being threatened by this, I was relieved. We sat down and decided that I would keep the store running and Marshall would be the technologist. He didn't aspire to my job, so there were no turf wars. We were both comfortable with the arrangement we had worked out.

Marshall came on board full time in January 1993 with the title of vice president. Ultimately he played a major role in the success of Sensors Unlimited. While I started the company, he was its technical brains and in many ways the driving force, along with a couple of other key people behind its success..

He also has both feet pretty much on the ground. A few years ago he and I were attending "The Einstein Lecture," an annual event sponsored by the Princeton Area Chamber of Commerce and a number of local corporations. The speakers are always Nobel Laureates. This particular year featured Princeton University psychology professor Daniel Kahneman, awarded the 2002 Nobel Prize for his research in the area of "affective forecasting" (integrating psychology and economics to form insights into human judgment in decision-making). The topic of his speech was happiness.

Just before the lecture began, Marshall turned to me and said, "If my father were here he would say 'Happiness? What's that? You have food on the table, money in your pocket and a steady job. What else do you need in life?'"

Marshall and I grew up in an era where people didn't see the need to define or pursue happiness. Today, people spend a lot of time worrying about it – I'm happy, I'm not so happy; what's my happiness quotient today? Like me, Marshall focuses on the basics: following your vision, doing your work, earning your keep.

Marshall is full of anecdotes and is never afraid to repeat the same story twice … or three times … or four. He often quotes sayings from his father, such as his definition of Pi: "The ratio of how long it actually takes to complete a job divided by the time the person first told you it would take."

We complement each other perfectly. To this day I can say that we have never had a serious argument or disagreement – and we've probably had lunch together at the Princetonian Diner well over 1000 times!

Processing guru

Sensors now had a couple of scientists and an administrative assistant. We needed to find an outstanding semiconductor processing guy. That describes Mike Lange perfectly.

I first met Mike in the early Epitaxx days. A headhunter had called and asked me to interview a young semiconductor engineer. I demurred, but he countered with "Just talk to him. What do you have to lose?" So I did.

Mike Lange walked in and proceeded to blow me away with his knowledge of semiconductor processing equipment and techniques. When we showed him our equipment, not only was he familiar with it, but he actually adjusted it so it worked better than ever. I was impressed.

But the headhunter wanted his finder's fee, something on the order of $5,000. That's really cheap by most standards, but it was a lot of money for us in those days. Again I demurred, but he persisted.

I think we finally settled on $1,500. I probably owe that headhunter a million dollars. Mike Lange has been a big contributor to my success for the last 20 years and continues to push the state of the art in wafer development to unheard-of levels. He is his own man and has his own way of doing things. But he invariably solves any wafer-processing problem he faces.

Mike had stayed behind when I left Epitaxx to start Sensors, but I kept my eye on him. A year later Epitaxx was succumbing to the inevitable bureaucracy of an acquired company. It wasn't the kind of atmosphere Mike thrives in, and he was definitely not happy. So Sensors was able to hire him away in early 1993, just in time for him to start working with Steve Forrest at Princeton University.

Modest and unassuming, Mike has a really sharp mind. One of my greatest sources of satisfaction was that he was able to get his master's degree in engineering from Princeton University under Steve's tutelage.

Symbiosis with Princeton

We also helped recruit Steve Forrest, even though we weren't the ones doing the hiring. You'll recall that Steve had moved from Bell Labs to USC, and from there made a huge contribution to Epitaxx as a consultant. He was about to be even more important to Sensors – even though he never worked for either company!

Right about the time I launched Sensors, Princeton University began looking for a Director for its new Photonics and OptoElectronics Materials (POEM) laboratory. (This lab is now part of PRISM, the multidisciplinary Princeton Institute for the Science and Technology of Materials.)

Once again opportunity was knocking. Steve Forrest was still at USC, but he would be perfect for the job – and how convenient it would be to have him sitting only 3 miles from my new startup!

Since I knew Stu Schwartz, the chairperson of the electrical engineering department at Princeton, I suggested that we meet with Steve the next time he was in town. Steve and I would have dinner, and Stu could join us for dessert. I met Steve at the Winepress in Kingston, NJ, which was a popular Italian restaurant housed in a Revolutionary War-era building just outside of Princeton. We had a nice meal, and when it was time for dessert, Stu arrived. I conveniently excused myself. Stu and Steve kept talking.

My introduction to the Winepress is worth a digression. In early 1984 I was interviewing a candidate for Epitaxx and we decided to meet at this restaurant (which I had never been to) because it was convenient. At the end of the meal we got the check, and I was informed that they didn't take credit cards. Embarrassed and flustered I asked the candidate if he had any cash – the two of us owed about the same amount. We came up with about $20 between us – not enough.

Finally the owner George Guadagno came over to me and said "Look … just sign the back of the check with your address. I'll send you a bill." He instantly created a loyal customer who held many corporate events and had many great meals at his restaurant over the years.

Fast forward a few months: Steve is at Princeton, in charge of POEM, with a ton of equipment but little technical help to set it up and run it. Sensors has about 5 employees to perform our SBIR contracts – including our fantastic process engineer Mike Lange – but no equipment.

It was a marriage made in heaven. Steve "borrowed" Mike to set up his lab and train students on the equipment. Sensors and Princeton made a deal to share R&D contracts. The upshot was that Mike became the processing guru at Princeton, and we got the right to use Princeton's equipment for research and to run our wafers through their lab.

Mike served as liaison between the two groups, and as we slowly prospered he got to build another semiconductor clean room, this time at Sensors.

Recruiting with bagels

We often had Princeton grad students as interns. This benefited both sides. The students could use the work they were doing as material for writing publishable papers, and gain experience that made them more employable. For our part we got bright, eager colleagues and possibly future employees at a very low cost.

An internship is how Sensors wound up hiring Chris Dries.

Once a week during the mid-1990s a bunch of us held an informal morning meeting at the Chesapeake Bagel store on Nassau Street, right across from the University. Steve, Marshall, Mike, and some of Steve's students would gather for a quick breakfast and talk about how their projects were going. It was a good way to keep up – and get to know each other.

Chris was one of the graduate students who regularly attended the breakfasts. He had worked on some of our joint SBIR programs. It was obvious to me that here was a guy with success in his future. He may not have been the top student in his class, but he had a hunger that elevated him above the crowd.

Bringing him into Sensors was one of the best moves I ever made. I wasn't the only one in the company who wanted him. Mike Lange knew him from the Princeton clean room. Marshall Cohen worked with him on SBIR contracts and later mentored him in the imaging business.

Chris was one of my "pushover" hires, just like Mike and Marshall. These were all people who wanted an opportunity to max out their technology skills, and saw Sensors as the place to do it.

My negotiations with Chris took just about as long as my original talks with Marshall: around five minutes. As I recall, Chris asked for a modest boost in salary, I readily agreed, we shook hands, and both of us felt like we had made a great deal. And we had.

Chris was our "golden boy," going from his PhD at Princeton in 1999, to research scientist, to multi-millionaire, to VP of R&D in just a few years time.

Hiring Pushovers

All of the best hires at Sensors were "pushovers." They wanted to sign on just as badly as I wanted to hire them. That was true of Marshall Cohen, Mike Lange, Chris Dries, Holly Dansbury, John Sudol, Bob Struthers – even Greg Olsen!

That's the way it should be. If you have to work too hard to hire someone, bend the salary rules, give extra bonuses, spend a lot of time convincing them to come … then it's probably not going to be a successful hire in the long run.

Take Bob Struthers, for instance. We must have gone through about seven sales managers at Sensors between 1992 and 1999 before we realized he was the right person – for us. I remember entertaining one potential VP sales, wining and dining him and his wife, showing them around Princeton, devoting an entire weekend, alleviating all his concerns, increasing the financial package, and finally getting a "…..well….OK….I guess I can join." Four months later he was out the door.

Ditto another "superstar" brought in with a $50,000 headhunter fee. More time (and money) lost. In fairness, while a couple of these sales people were duds, some of them were really OK. The timing just wasn't right. I mean, it's hard to sell *anything* when your product isn't fully developed, or keeps coming back for repairs.

When we finally realized Bob was the right guy, he'd been under our noses for quite a while. Marshall had known him from Princeton Applied Research. He was an industry veteran, known to many in the optoelectronics field, a staple at trade shows.

He had been hired almost as an afterthought. One of our sales VPs took Bob on to sort of beef up the selling staff. Then this VP and the next sales manager moved on, and we needed someone to lead sales. There was Bob. And what a gem he turned out to be!

Maybe the reason none of us recognized his talent earlier is his demeanor. Bob absolutely defines affability. He works to put you at ease and make you feel comfortable. But behind that relaxed exterior is a very sharp mind and a good basic understanding of technology.

His years of sales experience also make him a good mentor. When an over-zealous sales trainee would keep offering modifications of a product to a customer who had already agreed to buy, Bob would whisper in his ear "When you see the guy reaching for his wallet, shut up and write the order!!"

I've often thought of and used that wise advice in my own dealings. Most importantly, it came in handy a number of years later, when I was trying to sell Russian doctors on my being healthy enough for space flight. We'll get to that story in due course.

Another easy hire, Erick Argueta, came about in an unusual setting. In my early country and western dancing days, I used to frequent a place called Oakley's, near Princeton. One evening I was busy reviewing an SBIR proposal that was due the next day. I decided to bring it along with me to Oakley's, where I would have a quick dinner and then get back to working on it.

I sat at the bar and asked the bartender if he could get me something quick, as I was in a hurry. He went into the back, came out with a plate of food almost immediately, and asked "Do you like pasta?"

"Sure," I said, and gladly took the plate, impressed with his quick service. When I began to read my proposal while eating, he glanced over my shoulder and said "Hey, isn't that a JFET?" (This is a special type of transistor.) "Yes," I replied, "how did you know that?"

It turned out he had studied electrical engineering for two years but took some time off to go to Australia and bum around. He had just returned, and like many young people in their early twenties who were casting about for something to do, had taken a temporary job as a bartender. We chatted a bit more, I told him about my high-tech company, and I left, impressed.

We had a few more conversations over the following weeks. One night I mentioned to him that we were looking for a technician, and asked if he'd be interested. "Sure," he exclaimed, and we had another outstanding Sensors Unlimited staffer. Erick was a great find who wound up getting not only his BSEE with our support, but also an MBA from the Wharton School in Philadelphia

My point is that hiring is not an afterthought, not something that you rush to do when the need arises, or maybe even pay someone else to do. It is the most important part of any entrepreneur's career, and it is a 24/7/365 pursuit. You should *always* be on the lookout for good people: at conferences, baseball games, bars – you never know where you will find them.

Even if you are in no position to hire, keep your finds in the loop, keep them interested, make them aware that you have a great company. But always interview people, even if you can't hire them at the moment. How else will you locate good people? They are not always around when you need them badly.

Once, in one of our down periods, an employee came in and asked my "permission" to interview someone. I asked him why he was asking, and he replied sheepishly that he knew business was off and we probably wouldn't hire anyone. I said that might be the case, but there is always room for an exceptional person.

Everyone in the company can get into the act. Erick, Marshall, Chris, and many other Sensors employees loved it there and convinced their friends to come on board as well.

Performance enhancement

Mark Kasrel's contributions are a special case, because they're not about technology or sales or even company management. They're about getting the best people at Sensors to perform at an even higher level.

Though I'd spent a lot of time with the whiners and complainers at Epitaxx, I was determined not to do the same thing at Sensors. My time was better spent cultivating high achievers. For the most part I succeeded, and Mark had a lot to do with that.

Mark is a licensed professional counselor and business consultant who does management training and consulting with industrial companies. I've had an executive coaching relationship with him for over 15 years, and brought him in to help out at Sensors right at the beginning (1992-3). He continued to consult with the company until 2007, through three changes in ownership – he must have been doing something right!

At the very beginning, I told Mark that I wanted him to work with the high performers –to get them from good to excellent – and generally to avoid the slackers. I asked him not to be the "company shrink" – the guy to whom people went to with complaints, grievances, personal problems.

He readily agreed. We decided that he would politely listen to such people for five minutes and then firmly refer them on to others for help if he could. His main job was to help those who were already helping themselves.

Judging by the results he made a huge difference – and not just financially. To this day I believe Mark had a lot to do with making Sensors Unlimited a great company to work for.

Growing the business

For four years, from 1993 to 1997, Sensors was in a "grind-it-out" mode. We got by primarily on government R&D contracts, such as our SBIR funding, plus limited product sales.

Our government contracts were the ballast that kept the ship from flipping over and going down. We were employed, we were doing research, and we could pay the bills. But we weren't piling up money by any stretch of the imagination. It was a break-even business at best.

My personal finances during that time reflected the company's condition. Between 1993 and 1998 my net worth was effectively zero. The sale of Epitaxx had netted me a little over a million dollars, but every penny of that money was tied up in the guarantee for the bank loan I had used to start Sensors. I couldn't touch any of it.

Other people might have lost sleep over having risked all the money I had on a new company, with nothing to fall back on, but it never bothered me. If you asked me how I could tolerate that level of risk, my answer would be, "What risk?"

Somehow I never felt like I was taking risks. I've always believed that if things go bad, I can find another way to make a living – brush up on the electrician's skills I acquired from my summer jobs back in college and start wiring houses, or tend bar for a while, for example. Let me tell you, during the financial crisis of October 2008, this was more

Sensors Unlimited staff, 1994.

84

than a mere idle thought – the possibility became very real as we watched the stock market drop sometimes more than 10% in a single day!

Don't get me wrong. I love being able to afford the things I want – nice cars, vacation getaways, the apartment in New York. Believe me, I wouldn't want to go back to the lean years. Yet if I woke up one morning and found myself with zero dollars, I'd be annoyed, but I wouldn't take a gun to my head. Instead I'd say, "How do I get out of this mess?" And then I'd get to work.

Modest beginnings

There certainly were no luxuries in the early days at Sensors Unlimited. The old Epitaxx facilities we rented from Mr. Potts were the same low-rent space I remembered. We even got my old clean room back. Over the course of several years, with the landlord's blessing, we brought in more equipment. Ultimately we became self-sufficient, and no longer needed the POEM facility.

It was in that modest facility that we began the long road to profitability, starting with government contracts and some products

Most of our R&D contracts were from the SBIR program. This required most federal agencies, including NASA and the Department of Defense, to set aside a few percent of their research budgets for contracts with American-owned small businesses. A small business under the SBIR program is defined as having under 500 employees.

Even though we were continuing to work on sensor arrays, a lot of our product sales during the first eight years of our existence involved lasers – which we didn't make! This situation came about through my connections with my old employer, Sarnoff.

Sarnoff was no longer a corporate R&D lab. GE had bought RCA, Sarnoff's parent company, in 1987, and begun selling off divisions that

Sensors Unlimited staff, 2002.

didn't fit its business model. In 1988 it agreed to donate the labs to SRI International, a famous research institute in Menlo Park, CA, and the old RCA/Sarnoff Laboratories became an independent division of SRI that had to make its own way in the world.

Many of Sarnoff's stellar staff stayed on after the acquisition, including a lot of people I knew from my days as a researcher there. As one way of making a profit, the company had developed a business model of spinning off technologies into new venture companies. Unfortunately it was not producing the kind of returns they were hoping for.

So I reached out to Mike Ettenberg, my former colleague and boss, and the two of us quickly positioned Sensors as a different kind of vehicle to spin technology out of Sarnoff. Sensors essentially took on the role of packaging and marketing specialty semiconductor lasers manufactured in Sarnoff's lab.

Mike, of course, was a pioneer of the early days of semiconductor lasers, and headed up the laser group. Two of his scientists, John Connolly and Ray Martinelli, not only made significant contributions to the technology, they were real entrepreneurs. Or maybe I should say "intrapreneurs," a buzzword describing people who build businesses inside larger companies.

John and Ray championed new SBIR contracts, which we helped them fulfill. We started off as the sales and marketing arm for the laser group, and wound up helping them commercialize their technology.

Sarnoff manufactured the laser chips and we mounted and sold them to instrument makers. They went into devices used for a variety of industrial applications, including quality control and pollution control, based on sensing gases such as methane or carbon dioxide.

The devices were based on the principle that if you shine a laser beam through air that contains minute traces of gas, the laser beam gets partially absorbed by the gas at a certain wavelength. You can measure this absorption to identify the gas and calculate how much is in the air.

Just like the arrangement with POEM, both sides got benefits. Our employees had contractor badges to enter the Sarnoff buildings. They got to use support facilities such as the machine shop and stock room. Since we had access to a lot of expensive capital equipment, we didn't have to buy it ourselves.

We picked up other business as well. Marshall Cohen's relationships with Kadri Vural and Les Kosloski at his former employer, Rockwell, came in handy, and shows how companies in the same industry can collaborate to build a business, especially if they're not in direct competition.

While we were focused on near infrared, using InGaAs sensors, Rockwell was pioneering long-wave infrared sensors with a compound of mercury, cadmium, and telluride (HgCdTe). Once in a while, Rockwell would have a special need for a near infrared sensor and would come to us, or we'd need long-wave infrared capability and go to them for their sensors. We never had a formal agreement – we just trusted each other.

Minding the money

Through our partnerships with Princeton and Sarnoff we had the use of established facilities and support services, which held down capital investment and operating costs. This let Sensors live off the SBIR programs and limited product sales. As a result, for several years we avoided the need for external funding.

It also kept the bookkeeping pretty basic. Marshall and I sort of ran the company by ourselves for the first two years, including the financial side. I paid the bills on our "One-Write" checkbook, which left carbon copies behind as a record of expenses when you peeled off each check.

Talk about primitive! Today's simplest PC program for personal finances gives you more control over money than that old checkbook-based system, and we were using it to run a full-fledged company. (And by the way, how long has it been since you've seen a piece of carbon paper?) But we worked with what we had.

In addition to administering the company, of course, we had other things to do. Marshall and I both wrote research proposals. Government R&D was our main source of income, and we had to bring in projects. When we got a project, Marshall actually did most of the work, teaming with Mike Lange and Kevin Mykietyn, another former Epitaxx employee who had come along for the ride.

It was obvious that we would soon need help on the administrative side of the company. Bob Esposito, who had handled the Epitaxx transaction while at Peat Marwick, was a key player in the early days of Sensors when it came to finding people and services. He was especially good when stretching a dollar was required.

In early 1994 he introduced me to Holly Dansbury, who was then working as a Controller at another firm. Bob thought she might do well in an environment like ours, but it was clearly too early for someone of her caliber at Sensors.

Then things got busier, and the finances more complicated. In 1994 we tried to ease the pressure by switching to Peachtree accounting software, one of the early leaders in PC-based financial systems. But it was a lot harder to use than a One-Write checkbook. You almost had to be an accountant, not to mention a computer trouble-shooter.

I remember my assistant Jennifer Romano coming into my office in distress one day, saying "Greg … I just can't get this software to work properly…." I really appreciated the fact that she had come to me and told me of her difficulties, rather than just struggling in frustration. More people should take that attitude – it would make the workplace far more efficient.

Anyway, I told her "Don't worry … we'll figure something out," even though I didn't have a clue what to do. In one of the many "grace" events that have pervaded my life, the phone rang – literally about 10 minutes after Jennifer had left my office. It was Holly. Her company had just gone out of business and she was calling to see if we had any consulting work she might do.

I asked, "Have you ever used Peachtree accounting software?"

"Oh sure … use it all the time," she said. "Come right over!" I quickly responded.

In a flash, another great relationship was born. Holly Dansbury became our CFO and was a key member of our management team.

Financial adversity

We hired Holly not a moment too soon. Sensors sort of "ran a tab" at Sarnoff for the various services we used, and they billed us monthly. Money was often tight, and we didn't always have the ready cash to pay Sarnoff's bill when it was due. Holly would masterfully hold them off until we could get the money together. It didn't hurt that one of Sarnoff's accounts receivable people had worked for Holly in a previous job. That probably extended the credit a little farther.

After one of the many "dunning" sessions that grew out of this situation, I proposed that we give Sarnoff 10% of Sensors in return for a credit for $1 million of their services. Sarnoff was also pressed for cash, however, and turned me down.

That decision probably cost them a potential $50 million windfall a few years later, when we were acquired by Finisar.

In 1996 our business encountered the roughest patch yet. Sensors was in serious need of working capital. The only investment we'd ever received was that initial bank loan, guaranteed by me, which was now maxed out, and we couldn't live hand to mouth any more. So we went out looking for more funding.

I was never that great at impressing venture capitalists. Just my style, I guess. Whatever it is they're looking for, I don't have it. Words like "value proposition" and "business model" are as foreign to me as lofty spreadsheet-calculated five-year plans. Both Marshall and I are no-nonsense, grind-it-out types who think in terms of one year ahead and not much more.

Whenever I look back three years or more at almost any one of my ventures, I just smile and realize that I never could have predicted what lay ahead. It's amazing how much stock people are willing to put into a 3-5 year plan.

Marshall and I made several presentations to VCs but they always walked away. "Don't see the vision here …. where is your market…." To be honest, we didn't know either. But we had a blind faith that somehow, something good would come out of our foray into sensor arrays. I felt that way at Epitaxx too. And thank God it panned out, for us and for the many Sensors employees who did well.

Just to survive the downturn we had to take a number of belt-tightening measures, including pay cuts for employees. Upper management, which had the highest salaries, took the largest cuts, and hourly employees the smallest. But we usually avoided layoffs, and paid back all the cuts with interest when good times came.

Nobody was happy about the pay cuts when they happened, of course, but one engineer, Jeffrey Stratton, a recent hire, took them especially hard. Shortly after the cuts were announced he came to my office and announced he was going to look for another job. I didn't want to lose anyone with his skills, so I asked him why he was leaving a regular (though smaller) paycheck without another job in hand.

"It's not because of me – I'd stay if I could," he said. "But my wife wants me to take a steady job with a secure company."

It turned out that before joining Sensors he had worked at another small company that ran into such serious financial problems they'd stopped paying salaries. Management appealed to employees to stick it out, saying they expected to recover soon and everyone would be reimbursed. He'd believed them and stayed on.

After several months the company declared bankruptcy and closed its doors, stiffing all of its employees for their pay. His wife worried that Sensors was going down the same path. They'd been burned once – he had a family to support – she didn't want a repeat of that experience.

The guy actually left. Two months later we still hadn't been able to fill his spot. In one of my more inspirational moments, I decided to make another try at having him on board. What did I have to lose? It turned out he wasn't all that happy in his new job and really missed working for us.

To convince him to come back, I asked him, "What if I could guarantee that you wouldn't be taking a risk with us, even if we did go out of business?" I took out my personal checkbook and wrote a check to him for three months' salary, postdated to the end of the three months.

"I want you and your wife to trust me," I said, "so I'm going to trust you. If you come back here, and we don't make good on your salary in three months, you can cash this check. But if we do what I promise, I want you to give it back to me."

He stayed, and proved to be a great asset. He never had to cash that check. He gave it back.

Sensors in orbit

Marshall Cohen put it best: "We were a mousetrap waiting for the mouse to be invented." We had ideas why our technology might be useful, so we went looking for funding to develop these ideas, and for research business to support us.

Sensors gradually developed products based on our technologies. We maintained a subsistence existence based on three business areas: gov-

Inventing the mouse for our mousetrap: Marshall Cohen and I in 1994 with an award for best technical article from Photonics Spectra magazine.

ernment-funded R&D programs; the sale of linear InGaAs FPAs which we manufactured for a variety of commercial near infrared spectroscopy applications; and the sale of InGaAs-based cameras for commercial and defense imaging applications and lasers.

We developed new applications essentially by bouncing ideas off our market. One of the ways we did this was to write a lot of articles on the potential uses for IR arrays for trade publications like *Laser Focus World*, *Advanced Imaging*, or *Defense Week*.

The company didn't have money to advertise in those magazines, so that was our advertising. It also served as market research. Interested readers would come to us to discuss what we'd written, and some of them eventually turned into customers.

Laying the foundation

In 1998 Marshall and I made a bold decision to build a bigger, better clean room. It cost us $4 million, but with the growing interest in our technology, we were convinced we needed more production capacity. Although we had mostly weaned ourselves off Princeton's POEM facility, we were still dependent on a couple of operations there. In addition, our existing clean room was somewhat jury-rigged, and included a lot of old, used equipment. It was also limited to 2" wafer production.

Borrowing the money was a risky move. Sensors was barely above break-even, and servicing a $4 million debt would be a heavy burden, especially since the market for InGaAs chips was still in a developmental stage. Who knew how much it would grow?

Several factors justified the risk. First, we had just been awarded a $3 million contract to develop 4" diameter InP wafer technology by the Advanced Technology Program of the National Institute of Standards and Technology (NIST).

The money helped, but the nature of the project was the clincher. With 4" InP substrates we could get 42 chips from a single wafer, instead of being limited to the eight chips possible on a 2" production line. Making that many more chips in the same amount of time would give us much higher capacity and really cut production costs. We'd meet much less price resistance.

In addition, we wouldn't be paying a premium for cutting-edge equipment. GaAs wafers, used for cell phone and radar chips, were moving from 4" to 6" diameters, and silicon wafers for computer memory and processor chips were advancing from 8" to 12" diameters. The 4" wafer equipment we wanted was selling at a steep discount.

Our gut told us it was the right time to expand. We borrowed the money from a bank (the VCs still thought of us as a boring "niche" company) and Mike Lange got to build his new clean room.

It turned out to be a timely and fortuitous decision. Without our new facilities Sensors could never have capitalized on the demand for InGaAs arrays we faced just a few months later. Having a vision and following your dream sometimes pays off big. It did for us.

Market ignition

In late 1998 the fiber optics market started to heat up. Marshall had been discussing InGaAs linear arrays with research scientists at Bell Labs, which had spun out of AT&T as part of Lucent Technologies, the huge telecommunications equipment company. Suddenly the arrays were hot products. There was a feeling in the air that something good was happening.

All the buzz in the industry was around WDM, or "wavelength division multiplexing," a way to put the light from multiple lasers of slightly different colors on a single strand of fiber. This allowed one strand to carry as many as 80 different signals. A carrier such as AT&T or Verizon could greatly increase its capacity by adding WDM equipment to existing lines – no new lines needed.

The challenge with WDM is that the wavelengths of the lasers are very close together. If the lasers drift from age or temperature variations, they can start to interfere with each other. The industry needed a way to monitor the wavelengths and powers of the lasers at various points in the network to make sure this didn't happen.

This turned out to be the "killer app" for our sensor arrays. (A killer app is a new application for a technology that suddenly makes it a must-have item). Monitoring wavelengths is precisely what a spectrometer does, and we had founded our business on detectors for spectrometers in the near infrared band. In fact, the application looked very much like what Marshall had been originally trying to do at PAR.

Sensors began supplying InGaAs diode array detectors to Lucent Bell Labs, which was trying to make tiny near-infrared spectrometers that would be deployed across the fiber network. We also supplied the Lucent Fiberoptics division, selling first one, then five, then 10 units. Then they came back and asked, "Can you make thousands?"

We ended up supplying Lucent by the thousands. Then JDS Uniphase and others also developed spectrometers around our detectors. As the only company that could produce these detector arrays in volume, we quickly became the world's largest producer of InGaAs focal plane arrays in *any* infrared band.

Takeoff

Suddenly our new clean room was running flat out. After all those lean years, success felt great. We were looking to grow the company and get better at running it.

In early 1999 my old friend from Ridgefield Park, Rich Capalbo, visited Princeton. Over dinner he told me about a book he was writing – *The Significant Dozen* – about the 12 most important people in his life.

By coincidence we had about 12 top performers at Sensors – high achievers being coached by Mark Kasrel to be even better. I asked Rich if he would come and give a motivational talk to my people at the now-famous Winepress restaurant. It was a fantastic presentation. They all loved him.

Rich would stay on as a consultant. Little more than a year later he would handle the financial details of the wildly successful sale of Sensors Unlimited to Finisar.

Things were obviously moving fast, not just for us, but for anything related to fiber optic communications. Just to get some idea of the speed of developments, consider what happened to the Optical Fiber Conference (OFC), the trade show for the industry.

Historically the show had attracted a few hundred researchers. Then in 2000 attendance exploded. The OFC, held that year in Baltimore, drew 20,000 attendees.

Show coordinators were forced to abandon the standard policy of issuing badges and instead let entrants wear their business cards for identification. The exhibit halls were crawling with venture capitalists and analysts looking for the next big investment. That's how hot our field was.

Profit and loss

In just a couple of years Sensors had jumped from wondering where the next contract was coming from to doing about $20 million in sales. In 1999 we made a $3 million profit. In 2000 we made a $5 million profit that wiped out the debt from our clean room construction. We had nearly 70 employees, and started to think about the ultimate destiny of the company.

We had to do something to secure our future while we were successful. Startups do that by going public, which means raising money through a stock offering. Or they get themselves acquired by a larger, more established company.

We had always considered one of our products to be the company itself. As early as November 1999 Rich Capalbo had asked me over dinner to fantasize about what I would consider the "nirvana" price if we were to sell Sensors. I told him I would be very happy with $40 million. He asked Marshall Cohen the same question at a separate dinner and got the identical number. It was 2X projected revenues, which seemed reasonable.

In late 1999, when things were going very well for Sensors, I saw an ad in a trade magazine for the telecommunications marketing firm RHK and noted that they had brought on Jay Lebowitz, who used to work for me at Epitaxx. I shot him a short e-mail with congratulations and Jay mentioned me to his current boss, Peter Hankin.

Another coincidence: I knew Peter from the Epitaxx days when he had done a marketing study on us for Warburg Pincus. We exchanged a few e-mails and Peter suggested I go to a conference they were holding the next month at the Fairmont Hotel in San Francisco.

There I met John Soden from RHK. We got talking about what I thought Sensors could sell for, which evolved into "We have some companies who might be interested in acquiring you – like Lucent and Alcatel, as well as a Silicon Valley startup called Finisar."

So we set up a team to work on a possible sale. Rich was an MBA with lots of experience in mergers and acquisitions. John knew the fiber optics market. In March of 2000 we went to that suddenly huge OFC meeting in Baltimore and arranged to meet potential buyers.

Almost at once we were surrounded by a frenzy of suitors. Marshall and I did our dog-and-pony show (we had gotten pretty good at Power-Point presentations for financial people) for 12 interested companies.

This was near the peak of the Internet and telecommunications boom (soon to be relabeled a "bubble"), when big players were paying stratospheric prices for small companies with crucial technology. In the middle of our quest for a buyer, NSG sold my old company, Epitaxx, to JDS Uniphase for $400 million. The sale price was maybe 10X revenues. That made us sit up and take notice.

Docking maneuvers

Ultimately we talked with 13 prospective buyers, including Lucent, Alcatel, and Finisar – that's how hot our technology was. But we all felt it was a natural fit when we met with the co-founders of Finisar, Jerry Rawls (CEO) and Frank Levinson (CTO and coincidentally a University of Virginia graduate). It was a refreshing change to be talking to the top execs and founders of a company, both of whom not only understood technology, but were involved in creating it as well.

Finisar's market was mostly components for data communications within a building or complex. By contrast, we were the largest independent producer of products for long-haul telecommunications. Originally we thought that they were buying us so they could diversify into telecommunications. Actually, they wanted to use us as an in-house source of InGaAs detectors for their receiver products.

Flush with cash from a recent IPO, they still retained the fast-moving small startup culture. After three meetings with Lucent, we knew we'd have to sit through a dozen more before they made a decision. After three meetings with Finisar, we had a term sheet on the table.

As part of the decision process they came to Princeton in the summer of 2000 to see our operation. I had told anybody who'd listen that Sensors Unlimited was not just Greg Olsen and Marshall Cohen. They met our whole management team: Holly Dansbury (CFO), John Sudol (CTO), and Bob Struthers (VP Sales), as well as Chris Dries and Mike Lange, key technical people who later rose to higher positions.

Our building on Route 1 didn't look like much from the outside, but the inside was pretty impressive to any techie. We also took them to the Sarnoff Labs and the POEM facilities at Princeton.

Then there was an impressive lunch at the exclusive Prospect House restaurant on the campus of Princeton University (formerly Woodrow Wilson's residence when he was president of Princeton). Its dining room overlooks a beautifully manicured formal garden featuring a huge variety of decorative and flowering plants. What an atmosphere for negotiations!

By total coincidence John Ritter, head of patents for Princeton, came over during our lunch to tell me he had just learned that USC had granted us a license to use the avalanche photodiode Steve Forrest had developed there. Wow! It all came together.

That license gave us the technology to meet Finisar's primary goal for the acquisition: supplying their internal needs. All we had to do was develop and manufacture a line of high-speed PIN and avalanche photodiodes. We'd never done it before … but Chris Dries had done his PhD thesis on the subject! Sensors is probably still supplying them today.

As if things couldn't get any better, the three of us – Rich Capalbo,

Jerry Rawls and I – played nine holes of golf after lunch.

When the final offer came, it was stunning. Finisar would buy us for 20 million shares of their stock. At the time Finisar stock was selling for $30 a share (later that year it peaked near $50). We were getting $600 million for a company with $25 million in sales.

Some of the terms of the sale would come back to haunt Finisar. Half of the stock was distributed right away and could be sold after a short period of time. The rest would be retained by Finisar and issued a third at a time in three annual installments, provided Sensors met certain technical milestones.

On Rich Capalbo's recommendation we added a condition: Finisar would pay 9.5% interest on the retained stock, at a fixed value of $25 a share. Each year, when they distributed a third of the remainder, they would pay the interest in additional shares of stock at its current price. So we were continually accumulating more shares of Finisar.

Why not? Stock prices were going to climb forever, weren't they? The deal was announced on August 17, 2000.

Dutch uncle

I vividly remember that period, because I began to get cold feet about the sale of Sensors. Maybe it's like what some people go through on the eve of their wedding.

Here we were, with an offer for $600 million, which was growing by the hour as Finisar's stock price went up. What fool would quibble with such a fantastic deal? Well…this fool, I guess.

Companies were being sold left and right. Companies were raising money by issuing stock – IPOs (initial public offerings) were all the rage in the market. Maybe I felt like I was leaving money on the table…? It's hard to say.

As I think back on that two-day period, I get this sheepish feeling and say to myself, "How could you be so stupid?" But I almost was.

And once again I dodged a bullet. Two nights before we were going to announce our sale to Finisar I was having dinner with my friend and dance partner Carla. We were supposed to have dinner and practice a dance routine, but I remembered just sitting around like a mope and whining about how I was giving up my company, having second thoughts (maybe third), fantasizing about a $1 billion IPO, just sort of unhappy when I should have been elated.

At that point I decided to call the whole thing off as soon as I got into work the next day. I didn't sleep well that night, if at all. In the morning I called Rich, our deal maker extraordinaire, and told him I wanted to cancel the sale and explore an IPO.

Rich probably earned all of his fees and then some in the next 5 minutes. He said, "Greg, listen, whatever you want to do, I'll support you. Just do me one favor: don't tell anyone about this just yet. Wait until I can get a flight from California to New York before you do anything."

He and John Soden of RHK, who was in Dallas at the time, probably set the world speed record for getting to Princeton, NJ. They pulled strings to arrange flights. But by the time they arrived, I had calmed down quite a bit. Especially after my phone conversation with my longtime mentor Henry Kressel.

After my restless night I had called Henry to tell him what I was thinking. I got his voice mail, and left a message that said simply that I needed to talk to him. Henry was in Israel at the time, which is seven hours ahead of US eastern daylight time. But another Kressel trademark is that he will call you back any time, from anywhere.

It must have been late at night in Israel when he called, but I guess he could sense the distress in my voicemail and asked me "so what THE HELL'S going on …" in his usual diplomatic fashion. I told him what I had been thinking about.

If he had been in front of me he probably would have grabbed me by the collar and pushed me up against the wall as he spoke.

"Listen," he said. "I don't know where all these outrageous valuations for fiber optic companies are coming from in the buyouts and stock offerings." (He saw the coming crash before almost anyone else.) "But if someone is willing to pay $600 million for your company…you sign that deal without delay!!" No soft-peddling…no molly-coddling…no touchy-feely stuff. Just a plain gut reaction, which was so on the money.

By the time Rich and John arrived the next morning I had been beaten up by Henry and had settled down and came to my senses. I realized that this was a fabulous financial windfall for me and 100 other people. I was not going to mess it up.

As Rich and John walked in the door, I smiled and said, "Don't worry – I'm going to do the Finisar deal." I could tell by the relieved looks on their faces that they weren't pissed at all that they had made a completely unnecessary cross-country trip at the drop of a hat. It's all part of deal-making, which is their business.

After we had signed the term sheet with Finisar, another company came in and tried to offer us an all cash deal at a similar price. It was very tempting, but a deal is a deal. Unlike Wachovia Bank, which in 2008 agreed to be acquired by Citicorp, then backed out of the arrangement to go with Wells Fargo, we stuck with our original buyer.

Flying high

Once the acquisition had taken place on October 17, 2000, it seemed like Nirvana. We all believed that the fiber optics/telecom business would continue to grow at the rapid pace of the past year. We were rolling in dough, and orders kept pouring in.

To keep morale flying, Sensors instituted some free perks for our employees that rivaled those of Silicon Valley startups. For example, we had breakfast catered every morning. This wasn't purely altruistic, of course – the idea was to get people to come in to work early, since every extra hour of output was making us money. The free food was cleared away promptly at 8 a.m. – if you weren't in the facility by that time you missed out.

Our acquisition was proving to be an "accretive" one – that is, Finisar's stock price continued to climb *after* the transition, which meant that the market was judging it as a positive move. Everyone "knew" that stock options were worth more than cash, since the market for high-tech stocks, especially in fiber-optic related companies, kept going up and up.

This belief was thoroughly ingrained among Sensors staff too. Finisar had offered all of our employees the choice of a fixed cash offer for whatever Sensors stock they owned, equivalent to $25 per share; or the option of keeping their holdings after they were converted to Finisar stock. Not one person took the cash.

In my case, I couldn't take the cash. That's why one of the dumbest financial decisions I ever made was to join the Board of Directors at Finisar.

It seemed like a good thing to do at the time: you know, Greg could sort of keep an eye on things…feed the gang at Sensors the company gossip … everyone was all for it. Yeah, good idea until I thought about selling my stock!

In those days Finisar was doing an acquisition almost every week – or so it seemed. And anyone with inside information – like Board members – could not sell any Finisar stock while this was going on. This didn't bother me at the time of the Sensors acquisition, since we all believed that the stock would continue to climb and climb.

By late fall of 2000, though, high-tech stocks – including Finisar – were beginning to show hints of what was to come in the near future.

Values would drop, rise slightly, then drop again, each time never quite getting back up to their previous values.

My financial advisor, Jim McLaughlin, then with Merrill Lynch, is a thoughtful, cautious, non-confrontational type of person. But like Henry, he wasn't taken in by all the "this time it's different" hullabaloo around telcom/fiber optics stock. Jim had frequently mentioned the desirability of being diversified, and how it was dangerous for me to be completely concentrated in Finisar stock.

I wasn't worried because, hey, we all knew it would continue to rise! But Jim was making no such assumptions and constantly mentioned that as soon as a "window" opened up for me to sell some stock, I should do so.

Finisar's acquisition activity had me locked up until late December, when a 10-day window opened up for Finisar board members to sell. I remember the phone call distinctly: "Greg…this is Jim. We have an opportunity to sell…we should do it. NOW!". His normally conciliatory tone had been replaced by a sense of urgency that was hard for me to ignore.

I took his advice. We sold almost 90% of the stock that I was allowed to sell at quite good prices – prices that never appeared again. I was set for life, financially. I owe Jim big-time for that one.

I've learned to really value and trust long-term advisors like Jim and Stuart Tapper, my accountant for 25 years. Both are fiscal conservatives in the best sense of the term, and both have given me and my family members invaluable advice over the years.

Shortly after, of course, I got taken out to the woodshed by Jerry Rawls, Finisar CEO, for upsetting his major stockholders. "Do you know what this looks like," he said angrily, "a major shareholder selling most of his stock?" Jerry was just doing his job, but I was doing mine too. Subsequent events showed that I certainly did the right thing for me and my family.

Not every Sensors employee was as fortunate. More than a handful of potentially wealthy people ended up with little – or worse. It's one of my greatest regrets. But that's a story we'll revisit later. For me and a number of others Sensors Unlimited was not only a vision we were able to realize through hard work, persistence, and a few lucky breaks. The success we achieved allowed us to pursue other dreams.

Such as riding a rocket into space.

View from my NYC apartment in the Trump International Hotel and Tower.

6

Homesteading and space training
Progress and Reversals

A New Face in Space
An inventor plans to be the next civilian to blast off
Employees are whispering at the Princeton, N.J.-based Sensors Unlimited. CEO Gregory Olsen, typically a self-professed "nose to the grindstone" workaholic, has been disappearing for weeks at a time.... The last time Olsen acted this mysteriously was in 2000, at the height of the tech boom, when he sold the self-funded optical equipment firm... This time, Olsen's secret is really out of this world. On Monday [March 29] the 58-year-old research scientist, philanthropist and grandfather [announced] that he plans to become the next civilian to visit the International Space Station.

-- Brad Stone, Newsweek, 4/5/2004, p. 52

After selling my Finisar Stock in late 2000 I had the financial means to pursue any dream. I wasn't finished with Sensors or with being an entrepreneur – far from it, as we'll see – but there were some other interests I wanted to pursue.

So I embarked on one of the most tumultuous periods of my life. In just the next five years I bought a wine farm in South Africa, an apartment in New York City, and a ranch in Montana. My colleagues and I bought Sensors Unlimited back from Finisar, rebuilt it, and sold it again. For two of the five years I was also in hot pursuit of a seat on a Soyuz rocket.

None of it went smoothly. There were obstacles, frustrations, and disappointments at every turn. Some of the problems were more intractable than anything I'd faced while building Epitaxx and Sensors. Sometimes it seemed like luck had deserted me.

Everything came to a head in the fourth year, when the scrupulous doctors of the Russian space program booted me out of the Soyuz program – for the second time!

It was a crazy few years that began in December 2001, 18 months before I got bitten by the space bug, and just when I was beginning to enjoy the fruits of my financial windfall. To say things were not going well at Sensors would be a gross understatement – the telecom business collapsed that year – but we'll focus on personal endeavors, not business setbacks.

Seeking some personal space

Once I'd banked the proceeds from selling my Finisar stock, Mike Ettenberg, my buddy from Sarnoff, started hectoring me to get an apartment in Manhattan. He and his wife had just secured one of their own, and he was regaling me with tales of how great it was to have a place where you could spend the night when you went into the city.

I used to go into New York periodically for an opera, a show, dinner, or dancing. I countered that it was cheaper to reserve a room in the best hotel available. Mike's retort was, "But it's even better to have your own space. And you don't have to pack a bag for the weekend."

He had a point. I started thinking seriously about his suggestion and finally, in late 2001, decided to check out some apartments.

I distinctly remember the day I went looking for a place. It was Dec 4, and my Dad's old outfit, Local 3 of the Electrical Workers' union, was having a memorial service for the 17 electricians who died in the 9/11 attacks on the World Trade Center. I was going into New York to attend.

Since I would be in the city anyway, why not check out a few apartments while I was there? So I scheduled appointments at Essex House and Trump International.

My first stop was Essex House. I walked in wearing my black jacket emblazoned with "Local 3 Electrical Workers." The agent took one look at the jacket and disdain immediately registered on his face. He was probably trying to guess whether I was there to rent an apartment or to fix the wiring.

Clearly he thought I didn't belong there, and after a cursory breeze-through of two apartments, he asked if there was anything else that I wanted to see, as he had several other appointments that day. "No thanks," I replied, and walked out the door. I guess they didn't want my kind in Essex House.

From Essex House I meandered over to Trump International, wondering if more of the same treatment was in store for me there. However, my Trump agent, Susan James, a real fashion plate and a dyed-in-the-wool New Yorker, turned out to be a wonderful person too. After we made our introductions, she softly put her hand on my jacket and inquired if I was associated with the union. When I told her why I was wearing the jacket, her eyes misted over.

She didn't ask who I was or how much I was worth. She began to show me every available apartment in the building, always stopping to chat with residents. She knew everyone, and everyone seemed to like her.

Sometimes it takes just five minutes to know you're in the right place. Two months later, in February 2002, I signed a lease for an apartment in the Trump. My only regret is that I leased the unit instead of buying it – it's now worth about four times what it was selling for back then!

Long-term, though, I still wanted to buy a place, so I kept looking around. Susan had mentioned that a new building, the Time Warner Center, was going up across the street on the site of the old New York Coliseum.

It was a déjà-vu moment. This was the exact place where my fellow Fairleigh Dickinson engineering student Frank Palaia and I had attended that IEEE conference almost 35 years ago. The most valuable thing most college students got out of those events were freebie pens and pocket protectors. Frank and I got recruited into the PhD program at the University of Virginia.

I checked out some model apartments in the building in April. A unit on the 74th floor was still available. It sounded really cool: a view of all of lower Manhattan, Central Park, and the Empire State and Chrysler buildings – even the George Washington Bridge!

Although I was really enjoying my place at the Trump, in June 2002 I signed up to buy the apartment in Time Warner. It was a real kick to look out a window on the 39th floor of the Trump and see my future digs being built right across the street!

OK, so I'm buying this really cool upscale Manhattan luxury apartment – me, the son of an electrician from Ridgefield Park, New Jersey! I figured I needed some help in decorating the space, but I'm a pretty plain, straightforward guy. I knew I didn't want one of these "frou-frou" places where every square inch is covered with some knick-knack and the trendy furniture is featured in all the decorating magazines. Success has not changed that part of my personality.

For a while I struggled to find the right interior decorator. Finally I met this couple who seemed to understand who I was and what I wanted. I looked at pictures of places that they had done, which all seemed pretty simple and understated. Just what I was looking for. Or so I thought.

A tale of two continents

At the same time as I was acquiring my New York apartments I was also looking for homesteads. Having a place – or two – in Manhattan was great, but I've always liked the idea of a private retreat far from the crowds and noise of the city.

By the end of June I had two rustic retreats, in South Africa and Montana.

There were strong personal ties to both places. I've already mentioned how I'd come to love South Africa during my post-doctoral work at the University of Port Elizabeth in 1971. My daughter Kim spent her babyhood there.

Montana didn't carry the same personal connections, but an interest in the American West and especially General George Custer led me to explore and appreciate what that region had to offer.

Getting into wine

South Africa was my first plunge into serious land ownership. When I was there as a post-doc researcher, the country was an international pariah because of its *apartheid* system of racial segregation, which the white minority government used to maintain its control of the government. I went back to this beautiful land every five years or so to visit old friends like Hennie Snyman and Koos Vermaak, but never put down roots.

Then the country moved peacefully to black majority rule, under Nelson Mandela's African National Congress party. Its isolation was over.

In early February 2002, right around the time I was leasing my Trump International apartment, I flew there for some vacation, and possibly to look for a wine farm. Koos (short for Jakobus) Vermaak arranged for us to go around with a real estate agent and inspect some properties.

Koos and I had known each other for 30 years. As head of the Physics department at Port Elizabeth in 1972, he was my boss during my post-doc. In 1994 we were working together again when he joined the staff at Sensors, only this time our positions were reversed: he was working for me! Koos stayed at Sensors until his retirement in 2001, when he moved back to South Africa.

He probably doesn't know this, but Koos had a pronounced influence on me. We didn't have a lot of interaction while I was at UPE, but I noticed that Koos always seemed to be working together with other people instead of trying to go it alone. Hopefully I have incorporated his example into my own work and life.

A lot of the places on the market were wine farms (vineyards), because South Africa is a major wine producer, and we were in wine country. To give you an idea of the variety, there was even a kosher winery for sale! Because the South African rand had been devalued to an exchange rate of 13 to 1 against the dollar (it's seven to the dollar at this writing), it was a great time to buy. Everything looked cheap to dollar holders.

At each site I took notes and photos. On the long flight back to the U.S. I went over what I'd seen, trying to come to a decision. Within a week Koos called.

"I've found the perfect farm – it's just what you wanted," he said. "You have to see it right away." That was crazy, I told him. I couldn't just turn around and fly back to South Africa. He kept insisting, so I agreed to return in a couple of weeks.

He was right – it was beautiful. Later that month I bought my wine farm from a guy named Henny Andrews. It's a 400-acre spread in Paarl, just outside of Cape Town. There are 200 acres of vineyards and a three-bedroom house that dates back three centuries.

The farm had a foreman named Armand Botha, who lived there with his father Gawie (pronounced "Harvey"), his mother Herma, and a colored maidservant named Saarkjie. (Just so nobody thinks this is a racial slur, the term "colored" is actually an official and accepted South African designation for someone of mixed race.)

I retained Armand as foreman. I also used Gawie, a former game ranger who worked in heavy construction, to take care of the horses part-time, and Herma as a part-time bookkeeper to pay the bills and keep track of the financials.

But I needed a real manager to run the place. I was referred to a guy we'll call Jack, who was looking for something to do. It all seemed to fit, and I hired him.

With the operational side of the farm seemingly under control, I went back home and dove into running the business, buying the Time Warner apartment in NYC, and searching for a ranch in Montana.

Ranch dude

I had been going out to Montana since becoming fascinated by General George Armstrong Custer. When I couldn't sleep one night in early 1988, I started flipping TV channels to see if there was anything interesting on. At 3 AM this can be a frustrating quest. But then I stumbled on a documentary about Custer's Last Stand.

This was truly intriguing – I really hadn't known much about Custer before. In an odd coincidence, shortly after that a friend of mine gave me a copy of *Son of the Morning Star*, a biography of Custer that was a bestseller at the time.

My interest now piqued, I continued to read about the general and the Battle of the Little Bighorn, as Custer's Last Stand is also known. I was planning to attend a conference in Colorado in June, and thought maybe I would make a little side trip to the Little Bighorn Battlefield National Monument, which is located on the Crow Indian Reservation in Montana.

So I called the Battlefield, which is run by the National Park Service, and asked if they had official tour guides. They didn't, but suggested

Visiting my Montana ranch.

that I call Jim Court, the recently retired Superintendent of the battle-field, who now gave personalized tours.

When June arrived I spent 12 hours on the battlefield with Jim, and wound up having dinner with his family. We've been friends ever since. Jim, his family, and his infinite network of contacts really helped me find myself in Montana. I also got involved with the Custer Battlefield Preservation Committee, which buys and protects battlefield-related land from development.

I came to love Montana as much as South Africa, so it was no surprise that after the Finisar coup my thoughts turned to owning a ranch there too.

It's funny how my life experiences seem to keep repeating themselves. My purchase of the Montana ranch mirrored the process by which I wound up with the South African wine farm. In both cases a long-time friend showed me his home area … I fell in love with a property and bought it … management problems ensued.

I spent a day with Jim and yes, an agent, visiting three or four ranch properties, took notes and photos, and reviewed what I'd seen on the plane home, trying to make a decision. Once again I got a call from my friend a few weeks later, saying that I must come out and see this gorgeous place he'd found.

Of course the flight out to Montana is six hours, not 17, and there's no passport required! That made it easier. So out I went to view the old Shively ranch near Pryor, on the Crow Indian Reservation in southern Montana. It was all Jim had said and more. Beautifully nestled in the Pryor Mountains with an old weathered barn and horse corral from 1905, it is about four miles from the nearest neighbor, including the old Will James ranch.

I bought it in June, right around the time I was also purchasing the Time Warner apartment, and our management at Sensors was buying back our company from Finisar. I also wound up buying the Eagle's Nest Hunting and Fishing Lodge.

Pryor is a typical small town in Montana: a population of maybe a few hundred, a gas station, a post office, a small café and little else. It is about an hour from Billings, the closest place to get supplies.

It takes ten minutes' drive from town on the paved road plus seven miles along a dirt road to get to the ranch, "KristaKim," named after my daughters. During nice weather, you can do about 40 mph on the dirt road, which allows you to make it from Pryor in maybe 25 minutes or so.

But if it rains, or snows heavily…good luck trying to navigate that road!

To give you some idea of the difficulty, three years later, in December 2005, I was scheduled to give talks about space to school kids in Montana's capital city Helena (population only 33,000), and nearly didn't make it. I was staying at the ranch, 300 miles away, and it had snowed. When I woke up at 6 AM it was pitch dark and the wind had blown the snow into heavy drifts which blocked the road going out to Pryor.

My capable assistant Star had to take me out by snowmobile to meet our good Crow friend Wallace Red Star, who was waiting for me at the cleared end of the road. I've made some harrowing journeys, but that one ranks up there with the best of them. Yes, we made it.

During the first year or so I occasionally visited the ranch with Jim. There was an old 1950's style ranch building that had been used as bunkhouse during the spread's dude ranch days, so we had a place to stay. Other than that I didn't do much with the property. It needed a manager, but not right away.

Picking the Wrong Guy, Part 1

From June to December 2002 my attention was focused on getting Sensors back on the road to profitability. Throughout the year I would get glowing reports from Jack, my manager in South Africa, on how well

On patrol at the South African vineyard.

things on the wine farm were going, so I didn't worry about that at all.

It turned out these reports were often written by Armand. Looking back with the benefit of hindsight there were probably a few red flags, but I paid them no mind at the time.

At the end of the year I flew over to South Africa, looking forward to a nice relaxing vacation at the farm. First I visited Koos, staying at his place in Saint Francis Bay for several days. While there I phoned Jack several times but couldn't get through to him. That didn't worry me, because eventually I would meet up with him at the farm anyway.

Little did I know what lay in store for me. I drove from Koos' house to Paarl, but when I got to the farm, Jack was nowhere to be seen. My inquiries about him were met with several vague answers. Something just wasn't right.

Finally Herma, Armand's mother and the farm's part-time book-keeper, stepped forward to tell me I should be aware that Jack had a bit of a drinking problem. That was putting it mildly, as I subsequently learned. She was as diplomatic about it as she could be, but being the honest person that she was, wouldn't conceal the facts from me.

It turned out that Jack was getting so blasted during the workday that he would actually pass out on the floor and they would have to step over him. I asked Armand to drive with me over to where Jack lived, in a town about 30 miles away, so that I could sit down, talk to him, and see what was going on.

When we got to Jack's house we saw his car in the driveway and knocked on the door. There was no answer. Since the door and windows were open, we walked in and called out for him several times, again getting no answer. Maybe he was passed out drunk somewhere in the house? Eventually we walked upstairs to see if he was there.

Most of the bedroom doors were shut. We knocked on each, still getting no response. Finally I knocked on the door of the last bedroom and called out Jack's name again.

Something strange was going on. This door didn't sound like the others when I knocked. It was beginning to feel like we were in the movie "Psycho." After knocking a second time I reached for the doorknob and slowly turned it. There was a slight resistance. When I tried to push it open, it moved about an inch, then slammed back in my face.

I should point out here that Jack is a big guy, built like a bull, and he was standing on the other side of the door. I tried to reason with him through the barrier, telling him that I just wanted to talk to him so we could work things out together. But there was no response. Suddenly I realized how perilous the situation was.

"Well," I thought. "This is how people wind up dead. What an idiot I am to even think of coming up here." I motioned to Armand that we should both leave. But my mind was made up: Jack had to go, immediately.

Instead of having a nice vacation I had to go into fix-it mode. Back at the farm I started drafting Jack's letter of dismissal when I learned it wasn't just Jack. A man named Walter was staying at the farm too, as Jack's guest. Walter was supposedly a mechanic, but he was really a fellow alcoholic, and Jack had given him a place to stay so he could sleep off his drunks.

It was kind of creepy to have a guy like that around. He had to go too, right away. I approached Walter and courteously told him that Jack was no longer running the farm and he would have to leave. He seemed rather down about it, but agreed, and as I recall Gawie drove him into town.

A few hours later he was back. His lawyer had told him he had the right to stay on the farm. Off to his room he went. I discussed the situation briefly with Gawie and Armand, but when Herma told me that Walter was living on top of them and she didn't like it, I made up my mind that I'd get him out of there.

So I approached Walter and told him bluntly that he had two choices. He could accept 300 US dollars and leave, or we would evict him. Either way he was definitely going to be out of there that afternoon.

He thought about it for a second and decided to take the money. We drove him and his meager belongings to a homeless shelter and never saw him again.

Bad to worse

Unfortunately, the troubles kept coming. This was the day from hell, definitely on my list of the top 10 all-time worst.

As I pondered what I was going to do – how was I going to run a farm from 8000 miles away now that I had just fired the manager? – another crisis arose. The maid Saarkjie, who unbeknownst to anyone was pregnant, suddenly went into labor, and had to be rushed to the hospital to deliver her baby.

She had concealed her pregnancy out of desperation. Jobs are hard to come by for African women. I'm sure she was afraid that if we knew she was pregnant, we'd fire her and hire someone else.

As evening approached we all decided to have a barbecue. At least we could close out the day on a positive note. But by the time we put the meat on the fire it had gotten dark and the wind was starting to kick up. The smoke from the fire blew into the house, setting off the smoke alarm.

No one knew how to shut the alarm off, so it rang for 30 minutes. We finally figured out how to silence it and finished our barbecue. The day just was not getting any better.

By now it was close to midnight. I was hoping for a good night's sleep, even though I was jetlagged and had no sleeping tablets with me. As soon as we'd cleaned up after dinner and straightened up the house, I decided to go to bed.

I turned out all the lights and went into the guest bedroom, where I was staying. The whole place was in total darkness. I was feeling my way toward the bed and just as I got there my bare foot stepped on something soft and mushy and very smelly.

Jack had kept two dogs on the farm, unruly beasts that were barely housebroken. One of them had left a giant calling card on the floor, right beside my bed, which was now squeezed between my toes and forced into the rug.

I was irate. To make matters worse, for some reason the water in the house wasn't running. Fortunately there was a swimming pool, so I went outside, took a pail of water out of the pool, and at least was able to wash my foot. But with Saarkjie in the hospital, we had no one to clean the carpet.

At this point I was so enraged that I couldn't sleep. I just tossed and turned all night until daylight appeared. If Hollywood did a movie of the day it would have to be a comedy. At the time, however, it didn't seem funny at all.

Family-based management

When I finally dragged myself out of bed the next morning, I was so groggy that I decided to drive into Paarl to find a pharmacy or clinic where I could get a sleeping pill for the evening. Since I was in town anyway, I thought I might as well visit the hospital to see if Saarkjie's baby had arrived yet.

That was quite an experience. About 15 minutes of wandering around a very dingy hospital convinced me that if I ever got injured or sick, I'd find some way to get to Cape Town for care. Several corridors later, I finally stumbled upon Saarkjie's room. Her bed was empty. One of the other women in the room told me that she'd had her baby last night and had already left.

Then she went to the window and pointed to a phone booth. There stood Saarkjie, calling for a taxi to take her back to the farm. She was so afraid that someone else would fill her position while she was gone that only eight hours after giving birth she decided to check herself out.

I left the hospital and called out to her. I never saw anyone run so fast – and she had just given birth. We drove back to the farm where I finally sat down to breakfast with the Bothas, mulling over the situation and trying to figure out what to do.

Wolfing down Herma's always perfect farm breakfast, I worried about how I was going to run this farm in South Africa while I lived in the United States. I thought I had found the right guy to manage the place in Jack, and now I was back to square one. What should I do?

Top management: the Bothas, who run the wine farm in South Africa. L to r, Armand (Arrie) and wife Carin, Gawie, and Herma.

About every couple of years I seem to get a great flash of inspiration. Sitting there drinking coffee with the three members of the Botha family, all of a sudden the light bulb went on. Maybe the solution was right there in front of me.

Here were three very capable people who seemed well-suited to farm life. Armand already had several years experience at the farm, and his dad Gawie was an accomplished horseman and a former game ranger. With a mom who was a crackerjack bookkeeper, everything I needed was sitting at this table. This disaster might not be such a bad thing after all!

I grew excited as I thought of the possibilities and finally blurted out to Herma, "Would you and your family like to run this farm?"

They were surprised and a bit taken aback at my sudden request. But I was serious. I asked them to go back to their house to talk it over and let me know what they thought.

In the meantime I retreated to my house and started writing down my conditions for running the farm on a single sheet of paper, as I think I've done for most of the successful ventures in my life. About two hours later they came back to my house and said that they would love to run the farm for me. I showed them my list of conditions and they agreed.

We discussed money and tenant conditions for less than five minutes – the consistent characteristic of all my great associations. It proved once again that any time you find yourself in protracted negotiations over salary, conditions, bonuses due, it probably isn't going to work. Great hires are easy hires!

Fast forward six years, and I think that paper is still somewhere in a drawer. We've never had to look at it. They have been great managers of the place, marvelous stewards of the land, and putting them in charge has been one of the smartest moves I ever made. I have total confidence in their ability to run a farm, to deal with whatever problems arise, and to be great hosts to me and my guests when I visit the farm.

It's been a terrific six years. Once again, a near disaster had a wonderful outcome.

Wide Open Space(s)

Six months later, in June, the New York Times article on Space Adventures launched me on my quest to travel beyond the earth. An event at the KristaKim ranch in Montana a couple of months after that served to add to my enthusiasm about going to space.

On August 27, 2003 (coincidentally my oldest daughter Kim's birthday) Mars made one of its closest approaches to earth. It was a hot, sunny day – so hot that the power had failed in the New York City area. But I was out in Montana at the ranch, the perfect place for watching the "Mars-rise."

Once the heat of the day subsided I relaxed with a grilled steak and some good red wine. I remember watching the sky turn dark. Then out in the distance I thought I saw an automobile headlight approaching.

"Who the hell would be coming out here at this hour," I wondered. Or at any other time, for that matter. KristaKim is really secluded – there's not much traffic. So I watched attentively as this headlight seemed to approach. Then I realized it wasn't a car at all.

It was the planet Mars rising over the horizon in Big Sky Country! What a magnificent sight: the biggest, brightest thing I have ever seen in the sky, short of the moon and sun.

Two months later I was in Baikonur, watching a Soyuz roar away from the steppes into the big sky over Kazakhstan. Those two experiences cemented my commitment to embark on my own space adventure … as soon as I got past that little problem of being washed out because I'd had a collapsed lung.

Telling the World

As recounted earlier, I had the pleurodesis in February 2004 to forestall any possibility of another collapsed lung. Space Adventures believed that if I was going to get back on track for an October launch – my goal all along – I had to start training in April. That meant getting formal approval from the Russians very quickly. So in March of 2004 I went back to Russia with my medical reports, accompanied by Eric Anderson and Richard Jennings.

While I was hoping to get booked on the October 2004 flight, they were saying they could guarantee me for April 2005. And I still did not have a final contract with Space Adventures. I'd already put down money for their efforts thus far, and a contract would entail a second hefty deposit. There's no way I was going to do that until the Russians gave me something in writing that said I was medically okay to fly.

That's almost what we got. The agreement from the Russians specified that I was okay to fly as long as no other health problems emerged. They required periodic checkups to make sure nothing else had developed. If some other problem showed up, however, a kidney stone or a heart ailment or almost anything else, they reserved the right to reject me.

Still, I was elated. It was looking like a go situation. Space Adventures decided to make the contract public on March 29, 2004 with a New York press conference to announce that I would be the next private citizen to travel into space. Everything was beginning to move very fast.

During the week before the announcement Eric, Chris Faranetta, and I went to Washington, D.C., to do a little lobbying. I wanted to bring one of the Sensors Unlimited cameras onto Soyuz and take infrared pictures from space. The U.S. government was a major buyer of Sensors cameras and we knew they wouldn't be eager to give me an export license to take one into Russia, but I had to give it my best try.

We also wanted to make sure that NASA would not have any objections to me traveling with the Russians and being on the American side of the ISS once I got there. So we spoke with a few NASA officials and did the best we could to iron things out.

There were logistical issues too. Since I was going to use NASA's Internet, radio, and phone connections, we wanted to make sure those arrangements were in order. The trip to Washington was a good idea on Eric's part, as it smoothed out potential problems.

One of the people we spoke with was Representative Dana Rohrabacher, a 60ish hippie conservative Republican. His office looked like a California surf shop, decorated with surf boards and bathing suits. Apparently he had been quite a scuba diver and surfer in his time.

Rohrabacher was Chairman of the Subcommittee on Space and Aeronautics from 1997 to January 2005, and a big booster of private-sector participation in space travel. Eric went there to say "Hi, we want you to meet Greg Olsen, the next space participant."

We also had a meeting with John Marburg, the Presidential Science Advisor. We were covering as many angles as we could.

Public Relations Sensation

Space Adventures had retained a PR firm to give me some training in how to handle myself at a press conference: what to wear, what to say, what not to say, and most important, how to handle a potentially embarrassing question. They did a dry run where a PR guy fired questions at me, and got kind of nasty to see how I would react to a hostile reporter.

One thing the PR guy drilled me on was how to answer the obvious question: "Dr. Olsen, how can you justify spending twenty million dollars on a personal trip into space? Couldn't you make better use of that money by, say, donating it to a local charity or using it to help eradicate hunger in Africa?"

That's a legitimate point. My answer: "First, I have already donated more money than this to good causes, and I will continue to support those charities.

"Second, I'm going into space for some very good reasons. I'll be conducting several scientific experiments. I hope to share my experience with young people to promote interest in science and engineering. And I'm also going for the sheer joy of being there."

The press conference was scheduled for the trendy W hotel on 14th Street in Manhattan. Everyone was supposed to keep quiet about it until then, but I started revealing my plans on Saturday night, when I went to a local bar in Princeton and talked with friends and neighbors. One by one I told them what I was up to.

They were all as excited for me as I was for myself. "WOW – you're actually going into space? How cool is that!" I hadn't planned on it, but by now news was leaking out pretty fast. It was a good thing the press conference was early Monday morning!

By that time I was in sort of a heady "high," feeling really good about myself. After all the work and frustration, the dream was finally coming to fruition. The press turned out in force, and most of the questions were positive. Over the next few days I was interviewed by Harry Smith for the CBS morning show as well as by CNN, among others.

All the attention was certainly ego-flattering. To think that millions of people were watching me on TV! Maybe some acquaintances from my past were saying to themselves: "Hey, I remember that guy from high school."

The rest of the day was a whirlwind of activity. I was all over television – it was my 15 minutes of fame. I let myself enjoy it, knowing that in three months it would all be history.

From New York it was on to Japan for more PR, where I also met people like my old infrared camera rep, Nao Iwaki. He treated me like a hometown hero. Life can be pretty nice when everybody is patting you on the back and wishing you well.

As you might expect, the interview process in Japan was more formalized. All the press outlets came to see me at my hotel suite. First I met with one TV crew. They asked a bunch of questions. Then another crew came in and asked the same questions. Then a group of newspaper reporters appeared, asked the same questions, and took lots of photographs.

I don't remember anyone throwing me a curve. One Japanese reporter asked how much the trip into space cost, even though the figure had been widely reported. But it's not like he was trying to score points – he just seemed to want to hear me say "Twenty million dollars."

Picking the Wrong Guy: Parts 2 & 3

While all the negotiations and publicity for the space trip were going on, I was trying to wrap up a couple of loose ends: the décor for the Time Warner apartment, and a caretaker for KristaKim ranch. As if I needed more things to worry about....

But the situation was getting serious and time was short. Once I started training in Russia, I couldn't stay in close contact with the decorators in New York, or personally check up on the ranch in Montana. Arrangements had to be made before I left.

Setting Up the New York Space

The Time Warner apartment began to occupy more of my attention. Things seemed to go OK in the beginning. My decorators were quoting me prices that seemed sky-high but, hey, this was New York.

In their defense, they did brief me on everything they were thinking of doing and asked for approval beforehand. But by the time we started getting serious about what would actually go into the apartment, my eyes were focused on Russia, not Manhattan.

Anyone who knows me well can tell you that if something distracts me from what you're saying, you might as well be talking to a brick wall. To a large extent the problems I faced later were the result of my limited attention span.

One big red flag should have been when one of the decorators said to me, "Don't worry, Mr. Olsen...your friends are going to *love* this place." To hell with my friends – I'm the one who needs to like my apartment! But I didn't pick up on it.

Meanwhile, I was finding I really did love the Trump, as did my family and friends. It's probably the best place to live in New York City (maybe I should keep this secret to myself). The place is cozy and intimate, and you will never encounter a better, more helpful staff.

There were times when I wondered why I would ever want to leave, but then I would refocus on the fact that I was about to get a new apartment in what everyone was saying would be the best place to live in NYC. Each week a new floor went up. I could see my home being created right before my eyes!

In early 2004 activity around the space flight was reaching a crescendo. It was all I could do to keep from falling asleep as the decorators showed me fabrics and $16,000 end tables.

I do remember feeling a bit uneasy when they showed me some Greek-style columns that they wanted to use to separate the living and dining room areas. But I figured they knew what they were doing – what the hell did I know about NY apartments anyway?

My daughter Krista kept telling me that some of the prices were really off the wall … that she could find similar items at much lower cost. For example, they had found a custom furniture builder who built couches from scratch. Nice touch, but I began to wonder if I really needed this level of exclusivity.

I should have listened to my daughter and my own misgivings. Krista, who was overseeing the construction and furnishing of her new house, had bought her couches from a showroom in North Carolina and I loved sitting on them. She had developed a real expertise in décor, and she liked doing it. By contrast, I dreaded the whole process!

Looking back, it's clear I engaged the wrong people to create my new space. I made another mistake in letting them function as the general contractor as well. They hired the architect. Then they hired the contractor, who in turn hired the actual carpenters, electricians, etc. You can just imagine the finger-pointing when things would go wrong! Meanwhile everyone kept getting his/her fees.

By the time the building's structure was done and interiors were being built, I was on my way to start my space training. The apartment became an afterthought. I guess I just assumed everything would turn out OK.

Even so, once I got to Russia and started receiving e-mails on the apartment and photos of the furnishings, I began feeling uneasy about how the place would turn out. As the purchase orders started to bulk up my e-mail folders, I wondered again if all this was necessary. But I put these concerns aside and envisioned a neat NYC luxury apartment awaiting my triumphant return from space in late October 2004.

Cowboy heaven

The process of finding a manager for my Montana ranch bore an eerie similarity to what happened with the winery. Just as I'd found Jack through a referral, some of Jim Court's friends recommended "Rob" to me. I took a look at his resumé, and he seemed OK. I decided to interview him.

This was just after the pleurodesis, and I was home recuperating. For some reason Rob was in New York that weekend, so I invited him to my house in Princeton to meet and chat.

When I answered the doorbell, I was completely taken aback. There stood a guy looking every bit the Montana cowboy: jeans, western shirt, bandana, boots. He could have been auditioning for a Hollywood western. But he was in New Jersey.

I decided to ignore my initial shock and get on with the interview. As it turned out, my gut reaction was prescient. I should have listened to it, instead of charging blindly ahead.

Rob seemed friendly and had considerable experience with ranch matters, so I hired him. After all, I had this space trip looming in front of me and couldn't afford to spend a lot of time looking for a ranch manager. It was one less thing to worry about.

While I was in Russia for training everything at the ranch appeared to be going OK, although the bills were mounting up. Rob was sending me glowing reports (sound familiar?), and I basically signed off on everything he proposed.

Training Begins

On April 4, 2004, I began training in Star City, Russia, at the Gagarin Cosmonaut Training Center (GCTC). Life had undergone a radical change. Instead of running a business while pursuing various personal and family interests, I was going to concentrate exclusively on getting ready for the ride of my life.

There was a lot to learn. I had to deal with a completely new environment in Star City, which included picking up enough of the Russian language to survive training. There were also classes on the skills and knowledge needed for traveling into space, plus intense physical training.

And I was pretty much on my own. Eric Anderson of Space Adventures sent his sidekick Chris Faranetta along with me to serve as liaison, and Sergei Kostenko, head of Space Adventures in Moscow, was there to help out. But it was up to me to succeed or fail.

GCTC is located in the woods 35 miles northeast of Moscow. That would be home until the actual launch, hopefully in October, six months away. I live pretty simply, so I only packed a week's worth of underwear, socks, casual clothing, and sneakers, plus one suit (just in case I needed one). Of course my luggage also included some electronics, including my laptop, which I planned to take into space.

I also brought a few bottles of wine, but that was quickly consumed. Since I would be meeting lots of people there – and maybe host a party – I wanted to have wine from my South African wine farm. After I got settled Sergei arranged for me to meet the South African ambassador to the Soviet Union. The ambassador had attended the University of Moscow and spoke fluent Russian, very handy for getting things done there. I explained what I wanted.

He seemed a little surprised. "Wine? Is that all you want? I can get anything you need into this country," he bragged. But I said no, that won't be necessary, just the wine. So he helped me import five cases of wine and I made sure he had some samples.

The New Environment

Settling in was not a painless process. I was assigned an apartment in "Dome Chiteree" (Building Four), a typically dreary 12-story concrete-and-brick Soviet-era apartment house, built about 40 years earlier. My apartment was one of the three on the 10th floor.

It was fairly spacious by Russian standards. There were two bedrooms, a kitchen, and a pretty big living room. But there were downsides as well.

First, as in most apartment buildings I saw in Russia, the dimly lit hallway (there was just one bare bulb) smelled of garbage. In Russia the elevator shafts often double as garbage chutes. Your nose was reminded of that fact every time you used an elevator.

In addition, like a lot of Russian infrastructure, the building hadn't had much in the way of maintenance. The paint was peeling and many of the mechanical fixtures either didn't work or were heavily worn. They don't replace things in Russia very often. Instead they keep trying to fix them, or ignore the problems completely.

This is especially true of plumbing. The toilet in my apartment was typical – it was barely functioning, and always running. I was stuck with a lousy toilet.

Back home this would have been no big deal. I'm pretty handy and actually enjoy fixing things. I can usually fix any toilet or faucet within minutes, or maybe an hour if it requires a quick trip to Home Depot for tools or parts. But in Russia I had only a pair of pliers and a screwdriver, and there was no Home Depot.

Still, I tried. I even took the valve apart, only to find that the plastic seals were so worn and distorted that even the Rube Goldberg tricks I attempted didn't work. So I lived with the constant sound of running water.

That is, until one night when I was awakened at 3 A.M. by a woman pounding on my door and yelling for me to wake up. When I opened the door she just kept shouting "Vada! Vada!" That's "water" in Russian. I heard dripping in the bathroom and quickly went to my toilet, which apparently had been overflowing for some time. She lived below me; the water was pouring through her ceiling into her apartment.

I just kept repeating "isvenitsia" – "excuse me" or "sorry," a very useful word – turned off the supply valve, and started mopping up the mess while she watched. "Toileta nee robota" ("The toilet doesn't work") she said, which was pretty obvious.

I tried "isvenitsia, isvenitsia" again to soothe her anger, but she was clearly in no mood for courtesy. The next day I got the building supervisor to fix the toilet and left her a bottle of my wine, along with an apology note in my best broken Russian.

But I never heard from her again, which I thought was odd. I wondered what her story was, and how she came to live in Star City. While it is definitely a military base (you have to show proper identification to enter and to leave), I found out that many of the residents have nothing to do with the military or the Russian space program. They just happen to own apartments there, perhaps because a relative once worked for the space center.

I was pretty isolated at first. Chris Faranetta had an apartment a half a mile away. The Americans training for Soyuz missions were in the "NASA cottages," and I didn't get to interact much with them.

During the day I went to classes, and then to the gym. Mostly I ate at home, but occasionally Chris and I had dinner together at the Torgovy Center, a military PX-style store with a café on the second floor. The menu was very limited – Monday was chicken, Tuesday was beef stew, that sort of thing – but it was the only game in town.

We did get to know the waitresses (Tanya and Natasha) and they learned our peculiarities – that I wanted lots of butter to smear on their hearty Russian bread and that we didn't drink beer warm the way the Russians did. I don't think they knew that I was the "space tourist" (a term that I dislike but have grown accustomed to). To them, Chris and

Working out in Star City.

I were just a couple of Americans who probably had something to do with the U.S. space program.

Life in Russia

A couple of other observations may help make the highs and lows of Russian life more apparent. They center on civilized amenities and, once again, infrastructure.

First the bright spot. I turned 59 while I was in Star City. It turns out that Sergei, who's 19 years younger, has the same birth date. So we decided to celebrate together.

Like me, Sergei loves good food and wine, and he arranged for a meal at a great Italian restaurant in Moscow. Yes, there are great restaurants in Moscow. We dined together often and he never took me to a bad one.

I told him that he ought to get some little signs that say "Sergei Approved," and every time we went to a place he liked, he could leave a sign. He was so knowledgeable and perceptive that he would have had a great following as a restaurant critic.

But getting to those restaurants brings up one of the negatives of Russian life: driving conditions. First of all, there's the traffic. While Star City is only about 35 miles from Moscow, it's an hour-and-a-half drive. Imagine the gridlock trying to get out of midtown Manhattan at 5:15 p.m. and you'll have an idea of what the traffic is like between Star City and Moscow – all the time! The highway infrastructure just isn't there to support all this congestion.

To compound the problem, Russian drivers are maniacs. Just 15 years earlier the average Russian did not drive or even own a car, so the automobile is a new phenomenon for them. They are going through what I went through when, at 18 years old, I thought I was all macho because I had a "muscle car" and wanted everyone to pay attention, get out of my way, and appreciate its loud exhaust.

At the start of my stay my heart was in my mouth every time I got into a car. Eventually I got accustomed to the craziness, but I never got behind the wheel within the city of Moscow the whole time I was there. Fortunately we had a driver for that.

Lines of Communication

At first I did not find the Russian people overly friendly. Of course my almost non-existent Russian skills did not help. To make matters worse, learning Russian was a real struggle. I'm not at all like my daughter Kimberly, who can pick up a language with minimal effort (she speaks English, Spanish, Italian, French and German and can get along in several others).

I'd make mistakes on the most basic of phrases. One day I went to shop at the Torgovy Center after laboring through my English/Russian dictionary to find out the Russian words for all the things I needed. It was 7 P.M. when I got there. I looked at the clerk and blurted out "dobre utra" which means "good morning." She just chuckled at my mistake, which made me feel even more embarrassed.

One of my biggest frustrations was whenever I would attempt to use Russian (very few in Star City spoke English) the response was usually rapid-fire and unintelligible. It sort of intimidated me – funny, I can be very forceful, but things like that make me turn meek.

Yet I had no choice – I had to learn at least some rudimentary Russian if I was going to participate in the Soyuz program.

Thanks to Space Adventures, I was fortunate to have Inessa, a lovely Ukrainian-born woman, as my initial Russian language instructor. She intuitively understood my limitations and kept it simple and straightforward.

Instead of loading me up with a lot of grammar and corrections she went around the room pointing at objects and asking me to repeat their names. Bumaga, that's paper; karendash, that's pencil; and so on. It was a laborious process for me (and probably for her). But it worked.

Inessa is intelligent and broadly educated – she has a bachelor's degree in mathematics and a master's in music. She even taught piano. So many of the Russian people I met were bright and talented like Inessa that I'm tempted to say that the average Russian is smarter than the average American.

You could argue that my sample wasn't representative. I interacted mostly with the Russian elite. Okay, but in my experience their top seemed to be a little better than our top.

Even if they're not smarter, there's no question that they're better educated. Russian fifth grade math is far beyond what Americans are taught. When their kids come to the U.S. they usually do very well in school because they've covered the material already.

Anyway, even when you have a great talent pool to draw from, you have to spot the star performers when they walk in the door. I've always prided myself on my ability to find good people to work for me, but Eric Anderson might be a notch better at it than I am. He not only found Inessa, he found dozens of other really smart and effective people to work with Space Adventures.

Eric was so impressed with Inessa that he later married her. In 2004, when I was in Star City, they were married to other people, but those marriages both failed. They got together, and she left the employ of Space Adventures.

With Inessa's help I got steadily better in Russian. Being able to say something in the Russians' own language could help break the ice, but it wasn't the total answer.

For example, about a week after I arrived, I walked out of my apartment on my way to class just as my Russian neighbor emerged from his door. Eager to try out my Russian, I introduced myself with "*Otchen* [check], *preatna* [check]" ("Hello, glad to meet you"), and told him my name. He tentatively shook hands with me, mumbled his own name, turned around and went back into his apartment.

Granted there was a language barrier, but even taking that into account a lot of Russian civilians were standoffish. Rather than shy or unfriendly, they often seemed fearful.

This may have been generational. Before the reforms of Mikhail Gorbachev, who was Premier of the Soviet Union from 1988 to 1991, Russians were generally discouraged from speaking with foreigners. The over-40 set may be carrying that learned behavior into the 21st century.

When I did manage to get acquainted with some Russians, I found them warm and generous, with a great sense of humor. I struck up a small friendship with my "keylady," a term used for a woman who oversees things in Russian apartments, sort of like a "super" in New York. She was very patient with my limited Russian skills and replied in simple words which I often understood.

Washout

Suddenly it was the end of May. After two months of training at Star City, I had a week off over Memorial Day. It was prearranged that I would return to Princeton for a few days to attend the annual Sensors Unlimited offsite meeting. I'd give the employees a picture presentation on my experiences in Russia, and take part in our usual afternoon golf outing in the afternoon.

This was a Memorial Day tradition at Sensors. We'd been holding these outings since 1993. Since 1998 they had taken place at Jasna Polana, a luxury golf course with a fabulous mansion as its clubhouse and great facilities for business meetings and entertaining.

Being back in the U.S. also meant I could check out a developing dental problem and get some medical tests run.

Missing a molar and eight holes of golf

Quite early on the morning of the Sensors off-site I found myself sitting in a dentist's chair. I'd been experiencing some nagging pain in one of my rear molars, and was there to consult with my dentist, Dr. Howard Buchwald. He's an early riser like me, so a 7 a.m. appointment was no big deal – especially for a patient of 20 years' standing who had been training for a space mission in Russia.

Dr. Buchwald examined the tooth carefully and took x-rays, but found nothing wrong. He said it might have a hairline crack, which doesn't always show up on x-rays. Since the tooth was not critical to my dental health, he suggested that I might want to think about having it pulled.

I was having none of that. Off I went to my offsite meeting.

My Russia presentation went fine. It felt good to be home among old friends and co-workers, and I was moved by their genuine awe at my potential space flight.

After lunch we headed to the golf course for our annual "scramble" or "best ball" tournament. We played in foursomes. Each player would hit a ball and then everyone in the foursome would move their balls to match the best shot. This took the pressure off those who didn't play often or well and increased the camaraderie.

We had done this twice a year since 1993, on the Fridays before

Memorial and Labor Days. Very little work gets done on those Fridays anyway, and people often take off early for the long weekend or a vacation. My theory was, why not just make those days enjoyable for the employees? Mark Kasrel would do his "team building" exercises in the morning, and then people would either play golf or get an early Friday afternoon start on the weekend.

It's little things like this, and maybe a genuine statement of appreciation for your employees' efforts, that create loyalty and a positive attitude. You're also setting the stage for encouraging people to stay on during the hard times when salaries don't keep up with inflation, or to go that extra mile under difficult circumstances.

We didn't ignore the spouses, either, whose positive opinion is so important in keeping the workforce happy. We always held year-end Holiday parties where spouses also got to sample the luxury of Jasna Polana.

My foursome started out well, but then disaster struck. On the fifth hole, just as I was about to hit my shot, a fierce pain suddenly shot through my jaw. It was that tooth again! I could now feel a crack running right through it with my tongue. It must have been one of the lunchtime delicacies that finally opened it up.

I thought about persevering through the match, but Mark Kasrel quickly doused that idea. He insisted I call Dr. Buchwald, whom he also knew. I got the doctor on the phone and asked what I should do. Dr. Buchwald wasn't encouraging. The tooth would probably have to be pulled, which was serious business best left to a specialist. By now, he cautioned, most would have closed their offices for the holiday. He said he'd see what he could do, and that I should come right in. So I dropped my clubs and drove the six miles to his office.

Dr. Buchwald immediately saw that there was no choice: the tooth had to go. Fortunately he had found an oral surgeon who was still in her office, and he sent me over there for the extraction procedure.

She was already set up when I arrived. "So you're the astronaut…" she murmured as she prepared to administer Novocain. We chatted about my exploits. I would have used any ploy to ease the pain!

She was a real pro. In less than 15 minutes she had numbed me up good, extracted the tooth, and given it to me as a keepsake. She also gave me additional pain medicine and warned me to take it easy for the rest of the day.

Again, I was having none of it. My buddies were still at the golf course and my team needed me (or so I liked to think). Back to Jasna I went, catching up with my foursome at the 13th hole.

I remember the amazed looks on their faces when I showed them the huge molar: "You just had a tooth pulled and you're back on the golf course?" I don't remember whether we won or lost, but I do know that I finished the round and welcomed everyone else with a beer as they came off the 18th hole. It was an over-the-top implementation of my motto: Never Give Up!

That period, from April to early June 2004, was sort of a "golden" time for me. Training for a space mission at age 59? Not bad for a guy from Bay Ridge, Brooklyn and Ridgefield Park, NJ. I was feeling pretty good about myself, and the warm welcome I received on my Memorial Day visit to New Jersey only amplified the satisfaction.

Casual acquaintances seemed pleased to see me again (minor celebrity status never hurts), but my close friends, family, and co-workers were genuinely supportive of my striving to achieve my dream. I could feel their warmth and affection. My grandson Justin did his usual jump onto my shoulders and asked "Hey grandpa…are you really going into space on a rocket?" Yeah, I was feeling pretty high.

House party

The day after the outing my daughter Krista arranged a welcome-home party for me at her house in Colts Neck. She invited a lot of friends and family – more genuine affection, preceded by another great round of golf. I showed the slides of my Russian training and was peppered with questions.

Things don't get much better than that. Little did I know than in a scant 5 weeks I would be feeling much, much different….

Krista had just moved to her new house. It had taken nearly two years for her to oversee the construction of this dream house. During that time she, her husband, and their three young kids lived with me in Princeton. You might think this was inconvenient for me, even a big imposition, but I thoroughly enjoyed it.

First of all, I'm not a guy that hangs around the house. Weekdays I'm usually out the door by 8AM, and often not back until after 10. Weekends are taken up with golf and dinner with friends. Also I loved spending some early morning hours with my grandchildren, and enjoyed taking Justin for swimming lessons on Saturday mornings.

It was a great bonding experience. I get along really well with my daughter – I love to socialize and play golf with her – so there was very little conflict.

They had moved to Colts Neck while I was in Star City. It seemed strange to be alone in my house during that week back in Princeton. But there would soon be an even bigger change.

My house had some minor scuff marks on walls, doors, etc. and I had asked Krista to maybe get the house painted while I was away. Krista being a typical modern woman, this "paint job" escalated into a full renovation of my downstairs.

The work was supposed to start as soon as I returned to Russia. It seemed like a good idea at the time. Krista was an expert at design and renovation by then. But this plan was only to add to my misery when I came back from Moscow a month later.

Houston, do we have a problem?

As part of my week away from Star City I went to the Aerospace Medicine Division of the University of Texas in Houston for some tests, including a CT scan of my lungs. This fulfilled my agreement with the Russian Space Agency to get checked out periodically, and inform the Russians of the results.

I had been through so many tests that, like a horse being walked through its paces for the umpteenth time, I automatically did whatever I was told: breathe, don't breathe; don't move; walk, run, piss.

On the way back to Russia I had another appointment, this time in Farnborough, England, just outside London, where I was scheduled for a ride in a centrifuge. This was something I wanted to do to show the Russians that I was fit to go into space. It's a rigorous test involving multiple G forces at several different angles.

By the time I landed in England, the night before the test, it no longer seemed like such a good idea. I must have caught some kind of flu on the plane. I had a 104° fever and my head felt like it was on fire.

On the morning of the test I still felt woozy, definitely whacked out, even though I had slept for 10 hours. Luckily for me, the English didn't require a complete physical exam prior to the ride. A doctor listened to my heart and took my blood pressure; but did not take my temperature.

It is part of the astronaut /cosmonaut culture that you don't tell a doctor anything you don't have to. So, sick and all, I took the centrifuge up to 8 G's, round and round in this contraption about the size of a high school gym. I was in a capsule that could tilt to vary the direction of the

G forces on my body: right side up, sideways, left side up, upside down.

I spent 30 or 40 minutes in the centrifuge, with my EKG monitored the whole time. Surprisingly, I didn't feel like I was going to throw up, even though my body was ferociously tugged and pushed. I knew I had to tough it out. I hadn't come this far toward my journey into space to wimp out in a centrifuge. Failing a self-imposed test would look really bad to the Russians, and I didn't need that.

Insult added to injury

Another episode of that tough-it-out attitude played out in the Star City gym several days later. For the most part the trainers didn't stand over astronauts and cosmonauts (including me). You knew what to do, and you did it.

In my case I had a regimen of weightlifting and chin-ups. The chin-up bar was about nine feet off the ground – even with my height I had to jump up to grab on. There was a step stool for those who wanted one, but no one used it – it wasn't macho.

On that day I jumped for the bar with both hands – and my left hand slipped off. Instead of letting go with my right hand, I instinctively held on. Suddenly my right hand was supporting my full body weight.

I heard a muscle tear. No one else heard it or saw anything happen, but I was immediately in extreme pain. I got down, walked around, faked my way through a couple of routines (all leg exercises) for half an hour, and went home.

My arm continued to throb for the next couple of days, but I didn't tell anyone what happened. I couldn't. I still felt that my hold on the third seat on Soyuz was precarious – I would not be certain of a trip to the ISS until I was actually in the rocket.

Since I'd had other sports injuries, I knew that usually they got better by themselves, given time. I had time. It was June and I wasn't scheduled to fly until October. As it turned out, I didn't completely shake the injury for three years.

In mid-June Richard Jennings reported the results of my Houston tests to the Russians. It was a disaster. During the monthly meeting of the GMK (the Russian initials for Main Medical Commission, the panel that clears would-be cosmonauts for training and flight), the group was told that a black spot had shown up on the CT of my lung.

Dr. Yuri Voronkov, who handled my case for the IBMP and served as a liaison with other Russian medical organizations, told us my situation didn't look good. On June 20 I appeared in person at the GMK meeting. "We're sorry, Dr. Olsen," the panel said through a translator, "We hate to step on anyone's dreams, but we just cannot qualify you for space, and we have to reject you from the program."

I was devastated. The official letter that I got a few weeks later only made it worse: "You are barred from space flight for at least a year and maybe forever."

The next weekend was especially tough. My older daughter Kimberly was coming all the way from her home in Argentina to visit me in Russia. I had to tell her what had happened, and I couldn't hide my disappointment.

Despite my low mood we made the best we could of her visit. We went to the Bolshoi Theater and visited the Kremlin. I escorted Kimberly through the training site and took pictures of her sitting in the simulator. At that point most people in Star City didn't know that I had just been tossed out of the training program.

On the same day that Kimberly went back to Argentina, I left for Germany to have a biopsy done on my right lung. Getting drummed out of the program was bad enough, but what if that black spot on my lung was cancerous?

I guess I could have gone back to the U.S. for the biopsy, but Richard Jennings and I were hoping against hope that we could get a quick and encouraging reading from a German doctor. We did – the spot turned out to be non-malignant – but it was too late.

In retrospect (and I think Eric and Richard would agree) the whole fiasco might have been avoided if we had reacted faster to the report from Houston. There's an old rule in business that when you tell people bad news, you should also give them a solution. Never go to your boss and say, "I have a problem." Instead, say that you had a problem and here's how you are fixing or have fixed it.

We did exactly the reverse. We told the GMK about the problem without giving them a solution. If we had gotten a biopsy done as soon as we saw that black spot on the CT, we could have presented the film and the findings at the same time: "There is a noncancerous spot on his lung, but it has no effect on his health."

As distraught as I was with the GMK's decision, I never felt that there was any personal animosity toward me. Sure, I was angry, but it was anger at the system, not the people. No one there was out to get me.

Packing Up

Doctors have a lot of power over astronauts or cosmonauts or space travelers in any space program. They can say yes, you can go; and they can say no, you can't go. Virtually every astronaut I spoke to had medical issues at one point or another that could have nixed a mission for them, something as simple as a pulled muscle, the flu, or a cold. In a lot of cases they tried to conceal these ailments from the doctors.

But you can't hide something like a heart murmur or a spot on your lung.

Dr. Leroy Chiao, Ph.D., one of NASA's most experienced astronauts, was my cry-on-the-shoulder guy. I was supposed to fly with him and the cosmonaut Salizhan Sharapov in October. When I got the bad news from GMK I called Leroy up in the NASA cottages in Star City and said, "Oh, shit, do you know what just happened to me?"

He said, "Greg, why don't you come over and have a beer with me?" He'd had a similar experience, too, so he knew what I was going through. Leroy and I have remained good friends to this day.

Getting ready to leave Star City was very depressing. It felt like defeat, like I was giving up.

Sensors Unlimited infrared products to make the unseen visible: mini and micro cameras (top); image sensors and linear arrays (bottom).

7 Rebuilding a Dream: Sensors Unlimited

Firm resets its sensors for success
Greg Olsen sold Sensors Unlimited two years ago at the height of the technology boom, for nearly $700 million in stock. He just bought the Princeton company back – for about $6 million…. Now Olsen is rebuilding Sensors Unlimited. The company got its start with federal government research grants. He is going that route once again.

— Beth Fitzgerald, *Newark Star-Ledger*, 10/23/2002, p. 46

Washing out of the Soyuz space program was devastating – maybe the biggest disappointment of my life.

I'd had my share of setbacks over the years, and usually came out of them better than before. Having failed high school trigonometry I wound up at Fairleigh Dickinson University … where I got a great education. After grad school the only career option open to me was to take a job in South Africa, a pariah among nations at that time – and it turned out to be a totally positive experience!

My business dealings followed the same pattern. Epitaxx survived an intellectual property dispute with RCA plus a nearly disastrous run of product failures. Sensors Unlimited earned me enough money to underwrite my space adventure, but not before it had survived some difficult times of its own.

This was different. I had much less control over the situation. My chances of recovery ultimately depended on reversing my rejection by the Russian medical team. The only recourse was to tough it out, work with whatever resources I had, and hope for the best. There was still a chance that the Russians could be convinced to reconsider.

In a way this was a close parallel to what had happened with Sensors

Unlimited only a year or so earlier. Yes, Sensors had been acquired by Finisar in the fall of 2000, and it looked like we were on the road to sustained growth and profitability. But there was plenty of drama yet to come.

In fact, things began to go south almost before we were over the initial euphoria. Suddenly we were in a fight to keep the company alive that lasted right through the initial stages of my space training. The resolution of our problems proved that there is usually a way to salvage even the worst situation. It all related to the financial environment.

Sudden deflation

Flash back to October 2000, more than three years before my dismissal from the space program. Sensors Unlimited had closed on its acquisition by Finisar, and it was party time for all of us.

Stock prices were going up across the board. Everybody was sure they would continue to rise. Federal Reserve chairman Alan Greenspan's famous remark of four years earlier that the market was displaying "irrational exuberance" was a running joke.

We were hiring new staff during the acquisition to keep up with business growth. By the time the deal was done we'd grown from 70 to about 100 people, and most of them had Sensors stock. They were just as intoxicated by optimism – or exuberance – as the rest of the investing public. That's why they rejected Finisar's offer to buy their stock outright for the equivalent of $25 per share.

Instead they opted to hang onto their shares and have them converted to Finisar stock. After the acquisition, when Finisar's stock price continued to climb, this looked like a smart move. But it ultimately proved disastrous for many of our employees, as stock prices – and our business – started to unravel.

High finance, falling markets

A quick recap of the Finisar/Sensors deal will help explain how it eventually played out. Start with the big number: Finisar offered us 20 million shares of their stock, then valued at around $30 a share, for ownership of Sensors Unlimited.

Do the math: they were paying us $600 million worth of stock for our company. But also recall that we didn't get all that stock up front. Only half of it was paid out at the time of the transaction. Finisar was to issue the rest of it in three equal lots over the next three years.

It's worth going over this part of the deal again. Thanks to my buddy Rich Capalbo, our financial advisor, Finisar wasn't holding onto that stock for free. They had to pay 9.5% interest on it, and add enough extra shares to cover the interest at each distribution. These became known as the Capalbo shares.

There was an interesting wrinkle to this part of the deal. Although the interest was calculated on the basis of a fixed $25 per share value for the remaining stock, the Capalbo shares were to be issued at current value!

Suppose the stock price rose to, say, $50 at the time of a payout. Finisar would issue only 20 shares for each $1,000 in interest. This would be a good deal for them. On the other hand, if their stock went down to $10, they'd have to give us 100 shares – not so good. Ultimately the Capalbo shares proved so powerful that they determined our future.

Handling prosperity

Back to the big number: $600 million. A lot of people were amazed we'd gotten that much for Sensors. But some of our management actually thought Finisar got a bargain.

That's because the 20 million shares they put up to acquire us represented just 11% of Finisar's total value. Meanwhile Sensors was generating fully 20% of the revenues of the combined companies, and 78% of pre-tax earnings. Quite a deal – for Finisar.

Still, we couldn't complain. Almost all of our 100 employees were getting a windfall of $50,000 or more. Dozens got several hundred thousand, and maybe a half-dozen millionaires were created by this transaction. This was on top of my payout, which set me up for life and financed my space endeavor. Not bad for a bunch of lab rats who were just trying to make a living!

I knew that dealing with sudden prosperity can pose its own problems. Most people, including lab rats, aren't very sophisticated about finance. So during and immediately after the acquisition I made sure to provide our employees with as much financial information as possible.

We had several top-tier financial firms come in and give presentations about how to handle money, but in the end I thought it best for everyone to make their own decisions about whether to keep/buy stock, where to invest their money, etc. I did not want to be a "big brother" or seem paternalistic when it came to peoples' financial futures. It's a decision that has caused me some remorse.

There were a few people I offered to help directly. They usually replied with something like "Oh, thanks, but I have a cousin who is an accountant…he's an expert on all this stuff.…"

Yeah, a real expert. I still remember the woman who came into my office in tears when Finisar's stock price took a big drop, saying "I want to sell my stock but my financial advisor is telling me not to, to hold on until the stock goes up again.…" It never did – at least not to those high-flying values.

From boom to crash

But we're getting ahead of the story. Right after the acquisition, things couldn't have been better on the business side. Telecommunications was in orbit, and we could hardly meet the demand for our sensor arrays. Sensors was now a pure telecommunications provider, riding the wave of prosperity. Our new charter was to provide telecom chips to Finisar and do any R&D that was of interest to them.

My role had changed too. I was still president and CEO, but was arguably first among equals. Within the company I was providing leadership and being a father figure to the troops, as well as acting as overall cheerleader. Externally, given that we were getting a $600 million payout, my main concern was just making sure that the acquisition worked. I resolved in my mind to become a diplomat and help things go smoothly, at least on the Sensors side.

Sure, there were crazy corporate things creeping in. Finisar inflicted the highly complex Oracle financial package on us, great for big companies, but cumbersome for a 100-person small business. They also renewed our lease for seven years because of a tax saving gimmick. (This hurt us after the buyback when we were strapped for cash and didn't need all that space). But overall relations between the two entities were solid.

There was little I could do about these things anyway. I had been through an acquisition once before with Epitaxx, and knew that my days of being "the big enchilada" were numbered. The unavoidable bureaucracy of a larger corporation was taking hold.

None of this dampened our enthusiasm. With business booming, we continued to hire, increasing our staff to over 120. Most of these jobs were in anticipation of the "doubling" of sales that we all believed would occur, based on the optimism of our customers.

We had been supplying Lucent Technology with about 3,000 InGaAs linear arrays per month during most of 2000. They were using them in their channel monitors, which could keep track of up to 100 individual optical signals that had traveled over a glass fiber.

Since we were at about our maximum production capacity, it was often a strain to meet Lucent's monthly orders. I remember being anxious about their often-repeated concerns that our production facility was too small for their anticipated growth. They kept saying we should think about expanding or moving to a bigger plant.

Sensors also began supplying key components to our new parent company, but we were losing money on the work. While it could be argued that the losses were the result of setting transfer prices from Sensors to Finisar too low, the effect was to turn SUI into a loss-making subsidiary.

In the middle of all of this manic activity there was an ominous interruption. Around February of 2001, just five months after the euphoria of the acquisition, Lucent called us up and said "Hey…just this once…why don't you hold up shipment until next week?"

"Whew," we sighed in relief – finally, a chance to catch our breath! We could use the downtime. It gave us an opportunity to catch up on routine maintenance in the production facility and focus on non-production chores we'd been putting off doing.

A second week went by, then a month, with no further word from Lucent. Finally we began to suspect something was up – and it was. It turned out they had been storing our production for months in anticipation of orders from their customers. Those orders never came. Now they were stuck with an inventory of thousands of pieces of unsold goods.

Well, they never took delivery of another piece on our contract! Faced with the prospect of continuing to build a huge stockpile of InGaAs arrays they could not use, they really had no choice but to cut us off.

Yes, we had a purchase order from them. But just try enforcing a purchase order from a large American corporation. You'd have a better chance of winning the lottery! This is a lesson that all small companies should ponder well before staking a major part of their business on supplying big companies.

By mid-2001 we could see the writing on the wall. The big economic promise of the year 2000 was not to be. The fiber optics boom of 1999-2000 was about to become the fiber optics *bust* of 2001-2002.

Lucent's cancellation had a dramatic domino effect. We went from a profitable company to one that had to be fed money just to survive. All the easy money in the stock market was evaporating too, accompanied by a grinding crash in share prices. Finisar's stock price went off a cliff just like that of every other telecom company.

That's when reality hit our employees in a way they never antici-pated. Just nine months after they were celebrating their good fortune in owning so much valuable stock, they learned the wisdom of the old broker's adage: don't count your money until you sell your stock.

I had some very sorry people in my office, including big, grown men with tears in their eyes. They'd hung onto their stock when it was sink-ing, without understanding the tax consequences.

These folks had been warned about "short-term capital gains." They heard that the tax on money you made from your stock was much higher if you sold it immediately, rather than owning it for a year or more. Since they all "knew" that our stock would keep increasing in value, it didn't seem like much of a risk to hold it and avoid this problem.

For them a little knowledge was truly a dangerous thing. When prices fell through the floor, many of them wound up owing more in taxes on their stock than the shares were now worth! They found out too late that they were taxed on the price of their stock option when it was exercised (that is, when they actually got the stock), not on the greatly diminished value they got when they sold it.

It was a lesson for me, too. I should have insisted that all employees pay the minimum tax on their stocks on the day of acquisition. Rich Capalbo had strongly recommended this policy. But several employees wanted to avoid paying the tax up front, on the grounds that they would have the use of that money for almost 18 months before the tax would be due. I gave in, and lived to regret it.

As you'd expect, financially sophisticated employees made out well. But ill-informed ones, usually lower paid, were more likely to get hurt. If I am ever involved in a deal like this again, I will only accept cash for any employee whose share of the proceeds is worth less than a million dollars. Let the "sophisticated" folks play around with the stock market!

Telecom nightmare

In 2001 the whole fiber optics and telecom market simply evaporated. It wasn't just Lucent that canceled orders. Our sales dried up to almost nothing. I remember feeling slightly adrift at this time: no clear direction; no real purpose; just waiting around for the next corporate decision.

It came just after the tragic terrorist attacks of 9/11. With the whole country reeling from the shock of the disaster, our parent company forced Sensors to lay off about 30% of our staff.

We – and they – really didn't have much choice. Finisar had already laid off hundreds of their people, which was an even bigger percentage of their total staff. They had spared us in the initial rounds of cuts. Now it was our turn.

Nevertheless, I think Marshall Cohen would agree with me that laying off nearly 50 people – just about a year after our acquisition by Finisar had brought in all that money – was the worst experience of the whole Sensors Unlimited story. One of the hardest things I ever had to do in my life was to look people in the eye and say "Sorry...you no longer have a job...."

There I was, having cashed out for a big pile of money, riding high with millions in my pocket. I can't tell you how difficult it was to present a production worker who had shown up every day and worked hard with that kind of bad news. These were salaried people with kids in college, trying to make ends meet. Now they had to find some other way to support their families.

I can still see the looks of disbelief on their faces: what happened? How is this possible? There is no way to sugarcoat this process, no "user-friendly" approach. You just have to do it in a fair but final way, and feel lousy for a long time afterward.

To a certain extent I felt that I had made a Faustian bargain: a big financial windfall in exchange for a bunch of managerial headaches that I really didn't like. And now I had to look 40-odd working people in the eye and say "Sorry....but we have to lay you all off." This after the newspapers had been extolling me and the management team for our big win with Finisar.

I was rather ashamed of myself that day, and I felt genuinely sorry for all these people who were unemployed just nine months after they were enjoying lavish company parties, boat trips around Manhattan, and big Christmas bonuses. As I think about all the US banks that went belly up in 2008, and find fault with the executives who got millions in compensation while presiding over the collapse, I wonder if that's how our laid-off employees felt about us.

We spent the next year "on the dole," receiving monthly cash infusions from a company that had purchased us to supply chips they could no longer use. I have to say that through it all, Finisar treated us as honestly and as fairly as you can under those circumstances.

Buyback and recovery

We were obviously a drag on Finisar's fortunes. By mid-2002 the fiber optics industry was in a state of total depression. Nobody was making any money, including us. We were bleeding out cash at the rate of many millions per year. Finisar did not need our production capacity, and they certainly did not need our burn rate.

We felt constrained too. Being part of Finisar meant we couldn't do the military/industrial R&D and products that used to be our bread and butter, the projects that built our business.

We could have given up. Instead, we did what I believe you have to do in circumstances like these. We marshaled what resources we had and took on the challenge of rebuilding the company. As strange as it seems, our opportunity to rebuild came about almost by accident.

Every three months the Finisar board met in Sunnyvale, California. As a board member I attended every meeting. Before I went, I usually met with the Sensors management team to see if they had anything they wanted me to bring up at the meeting. That's how the idea of a buyback was born, thanks to a suggestion from John Sudol, our Chief Operating Officer.

John, by the way, was a great COO. A mechanical engineer by background, he had just completed an MBA at the Wharton School when we first encountered him during the early days of Sensors. He was looking for an interesting high tech opportunity, had heard about us, and inquired if there might be a position for him. He seemed like a solid guy, but that Wharton MBA, and the salary level that it usually commanded, probably scared me off a bit. Not to mention that our manufacturing was hardly ready for prime time!

But John is no quitter. He liked us and kept coming back. Finally we hired him, and it was one of the best moves we ever made. Nobody else in the company could have elevated our manufacturing to the production and quality levels we achieved during our golden years.

As I recall, in my meeting with Sensors management before the June 2002 Finisar board meeting, John sort of half-jokingly suggested that we do a "management buyout." His idea was to have Finisar buy out our employment contracts, so we could all vacation in Tahiti or some other idyllic spot.

Amazingly enough, John's offhand comment got me thinking in exactly the opposite direction. I decided to try to buy Sensors back from Finisar. It seemed like a great idea, and it was.

Capalbo shares close the deal

This is where the Capalbo shares came into full play. Finisar still owed us a third of the shares they had used to acquire Sensor in the first place. We were due to get half of that remainder as a second anniversary payment in October, along with interest in the form of additional shares. But the market had intervened since the original deal was made.

Finisar's stock price had plummeted to $0.60 per share. This had a perverse effect on their ownership situation. The number of Capalbo shares Finisar would have to transfer to us to cover the interest payment was so huge that it would have handed over control of much of the company to the former Sensors shareholders – meaning us. I would have wound up as the largest single owner of Finisar stock. No one wanted that, including me.

There was no simple way for Finisar to get around this quandary, either. If they did anything to avoid the distribution, such as firing the Sensors employees, it would cut off their supply of key InGaAs components, since we were the only ones who could make them.

It's not surprising, then, that when I brought up the buyback idea at the Board meeting, Finisar readily agreed to the concept. Somehow we came up with a purchase price of $6.1 million in cash. Sounds like an outrageous bargain – a steal at a penny on the dollar, compared to what Finisar had paid for Sensors just two years before! It only took about a month and a half to finalize everything – one of the fastest, easiest deals I've ever done.

Most of the Sensors senior management team chipped in money for the purchase price. In addition to the cash, we agreed to provide Finisar with a three-month inventory of key components, and to transfer production equipment and technology to Finisar employees so they could make the avalanche photodiodes they were so interested in when they bought us. (As far as I know, they never made a single one.)

These extras amounted to another $3 million in chips and equipment, and I had to put up that much cash as a security that we would deliver on our promise. But all in all, it was a good deal for both sides, as proven by the fact that the two companies maintained good relations long after the split.

To minimize tax problems, our management team formed a new company, hired all the Sensors employees, and then entered into a purchase agreement for the bulk of the subsidiary's assets, including the right to use the name "Sensors Unlimited." What was left of the subsidiary was moved to California. Finisar retained a 19% interest in the new Sensors.

Starting over

We had originally built Sensors Unlimited around supplying InGaAs photodiode arrays and cameras to the industrial spectroscopy and military markets, including process control. When telecommunications companies discovered that our photodiode arrays could be used to monitor optical networks, we had shifted our focus to concentrate exclusively on telecom products.

Now, with telecom in the tank, we had to diversify our market position to achieve profitability and long-term revenue growth. So we went back to our roots. Besides, we were really good at doing R&D and developing new products for the military and industrial sectors.

Times were tough. It wasn't clear that we were going to make it. But at least we felt in charge of our own destinies, and we were much nimbler than a larger corporation. Speaking for myself, I sort of welcomed the uncertainty of the future.

Shift in command

Right at the start of our new incarnation we appointed Marshall Cohen president. As I've said before, I consider Marshall a true co-founder of Sensors, not only because he was thinking along the same lines even before the company was founded, but because he slogged beside me right from the beginning, through good times and bad.

To give credit where it's due, there's another guy I think of as a co-founder: Steve Forrest. Even though he held university professorships throughout our history and never actually worked for the company, Steve was a big part of our success, serving as technical "superconsultant" for both Epitaxx and Sensors. Actually, to call Steve a consultant is almost an insult. He was with us from day one, and had a real commitment to our success. In addition, many of the technologies we used were originally developed in his laboratories.

Marshall and Steve are both brilliant guys, both Michigan grads, and work really well together. How bright are they? Well, Mike Ettenberg of Sarnoff and I joke that if they ever put both of them in a room and gave them IQ tests, we're not sure who would win. The only thing we are sure of is that they would not let either one of us into the room!

Another top exec in the reborn Sensors was Chris Dries. He was already a rising star, and we made him vice president of R&D.

I got "kicked upstairs" as CEO and Board Chairman. I was quite happy with this. Sensors pretty much operated as a meritocracy and I was glad to see the high performers rise to the top.

Building market confidence

Our management team knew the stakes were high. We were risking our own money on the business. Our employees also knew the odds were against us, and that their jobs depended on achieving our goal of rebuilding the company. Everyone had to play their part well if we were to succeed.

For that reason it was essential that everyone in the company was on the same page. We were always very open with our employees about the direction of the company, its finances, and any other information people wanted to have, with the exception of salaries and other personal data. Everyone knew everything.

As part of this initiative we put out a business plan that made our goals as clear as possible. This was nothing new. During my fourteen years at Sensors we always put a lot of effort into creating a solid, credible one-year operations plan. Any employee who wanted to see it could read a copy in the conference room.

Notice that it only covered a year. I'm not big on long-range business plans. I think five-year plans are fairy tales and three-year plans aren't much better. They're essentially works of fiction, written on the basis of pure guesswork about what the future holds.

I'm not saying your planning horizon can't be longer than twelve months. But when you begin thinking out any farther, you have to do it in more broad-brush terms.

It's true that there are certain specific decisions that require longer-term planning. For example, when we built our new clean room, we had to justify the effort and expense over the following three years, and forecast what products we would be selling throughout that period. But I think that's about the limit to reasonable conjecture.

Our goal was to transform ourselves from a fiber optics chip maker into a supplier of infrared imaging devices for the military and industrial market. In the wake of 9/11, with security concerns driving new growth in sensing devices, it was a reasonable direction to take.

We spent the next two years re-establishing our credibility as researchers and contract winners. Chris and Marshall took the lead here. My role was mostly reviewing their proposals. Holly and John, our CFO and COO respectively, really came into their own as well in the financial and production areas.

We also applied for grants through the Small Business Innovative Research program (SBIR), which had given me the contracts I used to start the company in the first place. SBIR grants are restricted to companies with fewer than 500 employees, and as part of Finisar we were over the limit. But now, as a much smaller company, we were once again eligible.

Our first year back on our own was a very tough period for Sensors, financially speaking. Between October 2002 and October 2003 we lost over $1 million. It wasn't certain that we would ever get back on track.

But Marshall was running the show about as well as it could be done. He, Chris Dries, and the rest of the crew just kept working at meeting their goals. It was that steady determination that really drove the progress of Sensors II. Once again, the secret of success came down to having good people in place and letting them do their jobs.

Soon the company established itself as a premier provider of image-sensing devices in the shortwave infrared range of the spectrum. Its products included cameras for military range-finding and remote sensing applications. Marshall calls it "the business of capturing the imperceptible and making it visible."

On the industrial side we had devices that let companies use light to determine the makeup of plastics and other products. In addition to these bread-and-butter applications, we were also open to exploring other interesting uses for our technology.

One fascinating application was art inspection. Museums and art historians were using Sensors cameras to look beneath the surface of famous paintings so scholars could study how artists altered their initial conceptions as they developed the finished piece.

For example, the Phillips Collection in Washington, DC used a Sensors camera to examine Renoir's *Luncheon of the Boating Party*. They were able to see behind the pigment of the paint, right down to Renoir's initial charcoal sketch, and observe the changes he made to the original conception that turned it into a masterpiece. Other galleries that made use of our cameras included Washington's National Gallery of Art, New York City's Metropolitan Museum of Art, and the Museum of Modern Art, also in New York.

By October 2004 Sensors Unlimited was profitable again, and sales and revenues were still growing. In just two years we had turned it from a drag on Finisar's fortunes into a thriving small technology firm.

I will admit to being largely a bystander, especially after the space opportunity arose in June 2003. The credit for the success of Sensors II really belongs to the team of Marshall as president and Chris as the vice president of research and development, plus the many dedicated Sensors employees. I was immersed in the effort to be admitted, and then readmitted to the Russian space program.

The Sensors saga comes to a close

With solid profits and rising sales, Sensors Unlimited was once more riding high, but our prospects were limited. We could stay profitable, but we would likely always be too small a company to make a major impact in our most important markets: military equipment, security, and surveillance.

If you want to land the most lucrative contracts in those fields you really need to partner with one of the "big boys" – Boeing, Rockwell, Northrup Grumman, and their ilk. It's a fact of life that the federal government feels most comfortable entrusting their largest projects to billion-dollar aerospace companies. Giving them to small firms is just too big a risk.

Knowing that, we began looking for a buyer for our reborn enterprise. While I would not claim to be the driver in the eventual buyout, I did participate in the selection of an "M&A" (mergers and acquisitions) firm to actually negotiate the deal. We began the process in early 2005, and talked to several top-notch M&A specialist firms.

At that point my financial advisor Jim McLaughlin, then with Merrill Lynch, suggested that I talk to one of their M&A people. I agreed out of courtesy – after all, Jim was the guy who guided me in selling my Finisar stock at just the right time – but I remember thinking, "What could somebody from Merrill Lynch, a stock brokerage, possibly know about the infrared sensing market?"

Quite a bit, as it turned out. I went into the meeting with Greg Starkins somewhat skeptical. In about 15 minutes he blew me away. He knew every military program there was and the names of influential people that we didn't know existed. He also had an amazing grasp of the technology.

I figured a financial guy with that much knowledge could probably swing us a good deal. So did the other Sensors people. We hired him and we never looked back.

I was in Russia by mid-2005 and wasn't even a participant in the actual negotiation with our eventual acquirer, Goodrich Corporation. By October 2005 Sensors had reached nearly $18 million in annual sales and $3 million in profits. The team had done more than turn it around – they'd made it attractive for someone else to purchase.

It was during that month, when I was literally up in space, that Sensors Unlimited negotiated its sale to Goodrich for $60 million. Rich Capalbo once again did a fine job with the negotiations. As Board Chairman I actually had to give Marshall my power of attorney to sign the deal. I credit the management team with that one: I just went along for the ride – and a mighty good one it was!

Goodrich was formerly known as BF Goodrich, a leading maker of tires for automobiles and other vehicles, including aircraft. In 1988 they had sold the tire business and the BF Goodrich trademark to focus on aerospace, and had since become a $6 billion player in that arena. They needed a night vision capability and Sensors filled the bill.

As a company Goodrich was a model of integrity during the whole negotiation process. Their negotiators made sure everyone was treated fairly. Sensors had found a good parent, and everyone made money and went home happy.

As for me, even before I came back from space, I knew that it was time for me to leave Sensors. Yeah, I still thought of it as my baby, but I would serve no useful purpose. I'd just be a figurehead, and that's not my style. I like to start new companies, do new projects.

Although I was no longer an employee after the acquisition, Goodrich generously let me use my office until year-end while I set up another office and my investment company, GHO Ventures. I was very happy that they also let me bring my assistant, Jennifer Kennedy, with me to the new place – and that she agreed to come. She has worked for me for over ten years. There have been more than a few times that her resourcefulness saved the day.

A couple of years later there were 60 people in the company and it was doing some $25 million. By that time Marshall had decided to leave, and Ed Hart from Goodrich took over as president. Chris Dries, the R&D leader, and Mike Lange, the technical whiz, also left to pursue their own interests.

Our outward migration wasn't because of anything Goodrich did. It's just the way we are all wired. When you join a big corporation, you're constrained in what you can do, and you're in competition with thousands of other people for middle and upper management jobs. You're no longer the big fish. You have to ask yourself, "Is this what I want to do?"

My colleagues and I didn't want the corporate environment. We liked being in entrepreneurial situations, exploring new directions, taking risks, making a difference. And we preferred the direct interaction of a smaller firm to the inevitable bureaucracy of a large organization.

The power of people and purpose

It's a funny thing, but at one time or another several people from the Sensors Unlimited team have implied or outright said to me that if it weren't for them, the company wouldn't have been so successful.

And you know what: *they're all absolutely right!* I'll take credit for putting Sensors together: incorporating, winning many of our initial SBIR contracts, putting in the early –and only – money, hiring the really key people, being the glue that held us together in the rough times. But it never would have happened the way it did without the other key players.

For instance, I'll never claim to be the technical brains behind Sensors. That distinction belongs to Marshall Cohen and Steve Forrest. And Chris Dries came on board as a young PhD from Princeton in 1999, made a huge impact through his technical competence and hard work, and probably could have run the company even without us.

But smart as Marshall, Steve, Chris, and many other Sensors people are, it wasn't overwhelming brilliance that got us to the promised land. It was hard work, perseverance, and the ability to work as a team and check our considerable egos at the door.

Overall, in virtually every area, the cast of characters we assembled at Sensors was the perfect combination. Mike Lange, Holly Dansbury, John Sudol, Bob Struthers all came along at the right time. They were all important factors in our success.

Take the Finisar acquisition, for example. It never would have happened with me alone. There is no doubt in my mind that what made us so attractive to Finisar was that we had assembled a great team, that the important role of each person was clear, and that we had supplemented internal resources with many important relationships with universities and research labs.

Visit the Sensors Unlimited operation as it exists within Goodrich today and you will find that each of these people left an indelible mark on the company, even though they are all long gone. We had a vision and a dream that we never gave up on, even when the market collapse threatened to take our dream down with it. I'm very proud of that.

Toward the end of the Sensors saga my dream of flying into space was also in danger of collapse. I don't think it was just a strange coincidence that its recovery coincided with the culmination of all our dreams for our company.

"Zero-G" training flight in Russia.

8 Rebuilding a Dream: Space Flight

Suiting up for spaceflight
The third tourist to visit the international space station ... could fly as early as October.... [Olsen's] trip, originally scheduled for April, was put on hold last summer because doctors in Russia found a health problem, which was not disclosed. In May, the Russian space agency gave Olsen medical clearance, and he resumed training.
— Mike Eckel, AP, in Philadelphia Inquirer, 7/7/2005, pp. B1, B4

It was the middle of 2004, a year before the sale of Sensors to Goodrich, and everything was going wrong.

Though space flight had become the center of my existence, it looked like I'd never get the chance. Thanks to a single spot on my lung I'd been blackballed from Soyuz training, maybe forever.

I was desperately searching for a way back into the Russian space program when by all rights I should have been helping to rebuild Sensors Unlimited. It was a good thing Marshall and the team were there to run the company.

In addition, there were some serious distractions in my personal life, mostly revolving around the Montana ranch and the apartment in New York.

All in all, the period between June 2004, when I was dismissed from the Russian space program, and April 2005, when it began to look like I'd get another shot, was one of the most "down" periods of my entire life.

Back to earth

My failure to finish training spread a pall of gloom over everything. After six days of medical tests in Germany, in the vain hope of convincing

149

the Russians to reverse their decision, I spent the long flight home totally depressed. I was dreading having to tell everyone that I had washed out.

Another rude shock was waiting for me in Princeton. Before leaving for Russia I'd asked my daughter Krista to get the place spruced up a little. I walked into my house, expecting to find it repainted with maybe some new kitchen cabinets. Instead it was in chaos.

There were bare wires hanging from electrical boxes. A lot of the sheetrock was torn off the walls. The downstairs had been gutted, the kitchen totally removed. It looked like a construction site...which is exactly what it was.

Krista assumed I'd be gone at least four months. (So did I, since the flight was scheduled for October.) With all that time to finish the job, she'd decided to expand the scope of the project into a substantial remodeling. It was supposed to be a welcome-back surprise. Of course when I walked in three months early, it wasn't ready.

Until the contractors finished, which took a couple more months, I had to camp out on the second floor. I couldn't even use the kitchen. My home was no longer my own.

To add to the general feeling of dejection, on July 1st, my very first full day back home, I would have to face up to the friends who had wished me well on my space adventure. It was my financial adviser and friend Jim McLaughlin's 50th birthday, and I was invited to his party.

This should have been a happy occasion, the kind of milestone you want to share with good friends. For me it was a devastating experience. Dozens of people came up to me and said, "Hey, Greg, we thought you were in training for space…." I wanted to crawl into a hole – for about 10 years!

Work provided no relief. Going in to my office at Sensors Unlimited meant repeating the same explanations to co-workers all over again.

To make matters worse, even though I had a business card that said Chairman and CEO, it was clear that my usefulness there had passed. Marshall, Chris, and the rest of their outstanding team were making the decisions and doing the heavy lifting.

My co-workers, especially the key managers, were super good about the whole thing. They mostly left me alone. Oh, they made sure to consult me on key decisions so I could feel like I was "in the loop." But I think deep down they understood what I was going through.

At best I was a figurehead. Realizing I didn't fit in any more, I was just brooding behind my desk and wondering, "Now what?"

Ups and downs

There were signs of trouble in other areas of my life, too – such as the Time Warner apartment. The decorators had been busy while I'd been gone.

I got an early glimpse of their progress in late July, which only served to reinforce my uneasiness about the direction they were taking. It looked like the finished space was going to turn out much fancier than I had envisioned.

In melodramatic terms, and with apologies to Thomas Paine, these were the times that tried men's souls! Or at least they tried mine. Here was Greg Olsen, eternal optimist and believer in working through problems, totally depressed!

I wish I could have just fast-forwarded the time from July 2004 until March 2005. It seemed to drag on interminably. Whenever I get down emotionally or have a bad day, I try to recall those days and the months that followed, and remind myself that things often get worse before they get better.

I turned to Dr. Murphy from Deborah Heart and Lung Center, who took X-rays of the lung with the ominous spot and ran numerous tests. I also went to the Johnson Space Center in Houston to get more tests. By September the spot was gone. It might have been a fungal growth that cleared up on its own. We never found out, and it never came back.

There were occasional meetings with Space Adventures and phone calls to consult with them. They were pushing my claims, but the response was generally negative. There wasn't much encouragement.

Flight insurance

Then there was the question of putting in a claim on my trip insurance. Yes, I had trip insurance.

Space Adventures requires that clients pay for their training and space flight in installments. Most of the payments had been made by June. If I couldn't fly for any reason, that money was gone. There was no provision for a refund.

As a precaution I had bought flight insurance from Lloyds of London. That way, if I tripped and broke my leg on the way out to the launch pad or had some other mishap that prevented me from flying, I'd get my money back. It was just like buying trip insurance for an expensive vacation.

Lloyds is used to insuring space flights – they provide coverage against failure for most communication satellite launches. But you know insurance companies: their first job is to collect the premiums. Their second job is to find reasons not to pay claims made on the policy.

While the last thing I wanted to do was admit defeat and try to claim the insurance money, my lawyer wisely advised me that I should make a claim within 30 days. If I didn't, I might lose any ability to make a claim later on.

So ... I reluctantly filed my claim. Lloyd's immediately came back with a request for copies of every x-ray, every medical report, every conceivable document related to my medical history. More grueling, boring, and depressing work. My heart wasn't in it. I wanted to fly, not get my money back.

As an aside, I'm not the only civilian participant to be washed out of the Soyuz program for medical reasons. Daisuke Enomoto, a Japanese entrepreneur, was also dismissed by the Russian doctors. He filed suit against Space Adventures in early 2008 in an attempt to get a refund. Maybe he didn't have insurance. I can understand his frustration, but wonder why he just doesn't get his problem fixed and fly.

Playing politics

How desperate was I to get another chance at space? Desperate enough to appeal to the president of the United States!

My financial advisor, Jim McLaughlin and I meeting President Bush.

I signed on to attend a July 2004 fund-raiser for the George Bush presidential campaign. Participation included a photo-op to get a picture with the president.

"Here's my chance to throw a Hail-Mary pass," I thought. Maybe the president could help me get back into the program.

As I stood on line, waiting for my 13 seconds next to President Bush, my mind was racing. What should I say – what should I do?

When my turn finally arrived, I was ushered up to the president, shook hands, and blurted out "Hey, Mr. President…I'm trying to go up into space with the Russian Space Agency but they won't let me…do you think you could help?"

He looked me square in the eye and said "…you're going up into space? Well, buddy, you have more guts than I do," and went on to the next person.

At least I tried, which is another variation of my mantra: don't give up – don't ever give up!

As I recall, Eric Anderson also attended that dinner. Although he had just celebrated his 30th birthday, he didn't seem very happy. I could tell it wasn't just business for him either – it was personal.

Getting some perspective

One of the things that kept me going throughout this period was a dim realization that although I was getting hit with crap left and right, my situation classified as a severe disappointment, not a total tragedy.

This was amply brought home one day in July, when the son of a good friend asked to meet me for dinner in Princeton. This guy had all the earmarks of a high achiever: good-looking, bright, and well-educated, with a great personality. Success was written all over him.

As we sat down over a glass of wine, I was feeling sorrier for myself than usual, and was preparing to unload my tale of woe on him. Then he informed me that he had lost his job. He'd been called in by his boss, expecting a raise and promotion, but instead was told that his position was being eliminated.

Shortly after that his wife had abruptly left him, and he was not going to be able to see his kids on a regular basis. Now here was real heartache!

As I went home that night I felt truly sorry for that young man – I would have prayed for him if I were religious. But I also felt grateful to him for making me realize the ultimate shallowness of my own self-absorption.

While this epiphany gave me a somewhat more balanced view of my situation, it didn't eliminate the depression. I still felt like I was dragging myself around, day by day.

One of the few bright spots came when I spoke on the phone with Dennis Tito, and then went to visit him. Dennis, the New Yorker who had been the first civilian to go into space, is a true pioneer and inspiration for all of us. He's also no wallflower.

His advice was: don't give up – do whatever it takes to make it happen – don't let anyone steal your dream…you can do it! It was exactly the encouragement I needed.

The Justin effect

Most of the things that were happening were downers in one way or another, but they were interspersed with some great memories – especially the times I spent with my grandson Justin, then three years old. I could be having the worst day of my life, and if Justin walked through the door and jumped up on my back (as he usually does), it would just light me up!

Justin learned about the planets from his Dad, and I tried to show him each one as it appeared at night. He would look out at the vast sky in wonder, a feeling I shared. I find stargazing awe-inspiring, knowing that the light reaching my eyes is hundreds or thousands of years old.

One night all of the planets that are visible to the naked eye appeared at once: Venus, Jupiter, Saturn, Mars, and Mercury. We made a special point of going out to see all five at the same time. I'm not sure he understood how rare this was, but both of us enjoyed doing it together!

During July 2004, in an attempt to get away from it all, I took Justin and his mom, my daughter Krista, for a stay at the ranch in Montana. It was still being run by Rob, the manager I'd hired before leaving for Russia.

When it was time to leave, we drove my SUV over the seven miles of bumpy dirt road between the ranch and the paved highway. As soon as we hit the pavement I knew I had a flat tire.

It was almost sundown and quickly getting dark. We didn't have a lot of time to change that tire. To make matters worse, I had to read the owner's manual before we could start.

That may come as a surprise if you haven't changed a flat recently. In the good old days swapping tires was easy: open the trunk, take out the jack and the spare, jack the car up, unbolt the flat tire, bolt on the spare.

Well, it ain't like that no more, especially with SUVs. I'm an engineer and handyman, but it's difficult for me to see how I could have changed that tire without reading the instructions first. I had to learn about tools that were unique to that car, and how to use them to perform some fairly complicated procedures.

Anyway, there I was, reading the manual in the dimming dusk, when Justin came and looked over my shoulder. He wanted to help. So...as I slowly figured out the instructions, Justin was with me every step of the way.

I cracked the nuts loose – he removed them. I got the spare from under the car (not an easy task) – he rolled it up to the wheel. I lifted it onto the hub – he put the nuts on and got them almost tight. I let him release the jack and we both looked at the finished job with a great deal of satisfaction.

Justin and I had a blast together. To this day we both still talk about it with glee, and Justin always checks my tires whenever I visit. For me, it is these small, special moments that make life truly worthwhile.

Homestead matters

Little by little, depressed and frustrated, I started to focus on other aspects of my life, like the ranch. I was starting to get a bit uneasy about some of the ideas that Rob was throwing my way.

Take the access road, for example. Admittedly nearly seven miles of dirt road is tough driving in the best of weather. After snow or heavy rain: fuggedaboutit! Night driving is no picnic either. Rob proposed that we pave the whole seven miles.

Montana friends: Jim Court, Wallace Red Star, and my ranch manager Baerbel "Star" Stuetzle.

155

I was having none of it. Why locate in the middle of nowhere in Montana to get away from people, then make it easier for them to get to you?

We differed on a number of other issues too, and it became clear – to both of us – that this wasn't working. One day he came to me and suggested we part ways.

I was so relieved! Although he did some things upon leaving that really ticked me off, like draining the gas tanks dry and taking my own personal hand tools with him, I was glad to be rid of him. But once again I needed to find somebody to watch the place while I wasn't there.

Fortunately my friend Jim Court stepped in and filled the gap. Since I hadn't started any new construction, there wasn't much to look after on a daily basis, so it wasn't a full-time commitment for him

While that meant I didn't need an immediate replacement for ranch manager, eventually I would have to find someone. Then another one of these "grace" things occurred.

About a year before this I had met Star, whom I've mentioned before, at the nearby buffalo ranch where she was working. That was our only contact, but I remembered being favorably impressed by her. Jim happened to run into her one day in Billings and asked her what she was doing. She replied that things hadn't gone too well with the buffalo ranch and she was now working as a wrangler in Yellowstone Park.

When he relayed this story to me, my mind raced with possibilities. Could lightning strike twice? Could a bad experience with a property manager be followed by a lucky break, like the one I'd had in South Africa?

I wasn't about to sit on this one. I asked Jim to make an offer. Within days he had tracked her down again and told her I was interested in having her live on my ranch and run it. To my relief, she accepted.

Now, some four years later, I rank her right up there with the Bothas in South Africa in terms of integrity, character, and competence. She is just a delightful person to run a ranch with.

Interior spaces

As the dog days of summer turned into autumn, I was getting increasingly uneasy with the choices the decorators were making for the apartment in the Time Warner building. By now I had visions of this fashionable space that just wasn't me.

Each week my stomach would churn as I envisioned living in the type of apartment that you see featured in newspapers and architectural magazines, filled with all sorts of knick-knacks and such.

When I saw the actual "Greek columns" that the decorators planned to install as room dividers – they looked to be about two feet in diameter – I finally rebelled. They agreed to use smaller columns. Subsequently, after moving into the place, I had them all removed.

I remember anticipating moving-in day in November 2004 with something approaching trepidation. I was afraid of what I would find. The decorators begged me not to go into the apartment until they had *everything* in place. I reluctantly agreed. Finally, the big day arrived....

My immediate reaction on entering the finished apartment was dismay and regret. Why had I let things go so far? Why hadn't I stopped the whole process months before? It just wasn't my style at all.

For example, there was a painting valued at nearly $100,000 hanging on the wall. I positively disliked it. (Fortunately, it was purchased on consignment – as were many of the more "frou-frou" items. I returned every one of them.)

And then there were all those high-end furnishings, the ones Krista had thought were horribly overpriced. I wasn't happy with any of them.

All in all, the apartment was another big emotional letdown to add to the list of woes. Kicked out of the space program – my house in shambles – no progress with the Russians – the insurance claim – and now this!

The decorators sensed my dissatisfaction right away. They asked me not to judge the place by my first reaction, but to sleep on it and see how I felt when I came back the next day. Not bad advice, and I agreed.

On the second day they could tell that I was truly disappointed. I think they felt just as bad as I did, and wanted to make good. They offered to do anything from completely redoing the place to helping me sell it. At that point I just wanted to be left alone.

Moving Monet

In hindsight I can't completely blame them for what they did. They were creating the kind of luxury apartment many wealthy Manhattanites would want.

But they didn't read me the way I think a professional service should read a client. If they had picked up on my responses a little better, they would have sensed that this New Jersey-bred son of a Brooklyn electri-

cian didn't really fancy a lot of ornamentation. I neither appreciated it
nor wanted it.

For me the real "pièce de resistance" came when I asked about bringing in my beloved statue of Monet's "Olympia" by Seward Johnson. He's
my favorite sculptor, and I've come to know him as a really nice man as
well. In addition to his original creative work on modern subjects, he
makes three-dimensional sculptures of classic paintings, mostly Impres-
sionist.

He had completely won me over with his recreation of Renoir's
"Luncheon of the Boating Party." That painting held special interest for
me, since a Sensors Unlimited camera had been used to disclose earlier
versions of the work under the final paint.

Naturally I was excited at the prospect of locating my very own Seward Johnson sculpture right in the corner of my new bedroom. It was
the perfect place for it. More important, it was my idea, not some interior designer's.

I had asked my decorators to visit Grounds for Sculpture in Hamilton, NJ, just outside of Trenton, and measure the statue. (Grounds for
Sculpture is a 35-acre public park founded by Seward Johnson to promote an appreciation for modern sculpture, and Olympia was on display
there.)

But my request had been overlooked, with the result that the deco-

Seward Johnson sculpture of Monet's "Olympia" in my New York apartment.

rators had not taken the size of the statue into account when laying out the interior. Now it would not fit through any of three doorways that led to my bedroom. The apartment would literally have to be gutted again just to put Olympia in her proper place.

This was deflating and depressing, all right. But being me, I immediately went looking for other, more efficient ways to move the statue to its designated spot.

Just tearing out the doorways didn't look like it would work. So I asked building management if we could put a crane on the roof and lower the sculpture from there. We'd take out a window, and pull Olympia through the opening right into the bedroom.

I'm familiar enough with construction to know that this is a very doable approach. Expensive, but doable. As a portent of things to come, management refused to allow it, citing insurance, safety, and other concerns.

It's only money

So here I was, ensconced in this "luxury" apartment that I really didn't like, depressed at how my dreams had once again been thwarted, and not really knowing what I wanted to do. With space still looking out of reach, I was beginning to feel like nothing I did would turn out right.

One day I was particularly down. I remember looking out the window at my old Trump Tower digs and wishing I was back there. So…I called Susan James, by now an old friend, and caught her in a rare, idle moment in her office at the Trump. She agreed to come over and check out my new apartment.

When she arrived, I proceeded to spill my guts to her. To this day I can hear her saying, "Greg, you know, it's only money…it's stuff…apartments and furnishings can be bought and sold…it's not life…."

What a prescient remark! If you don't like something, sell it and buy something you do. Obvious as that sounds, I'd been so mired in disappointment, I hadn't fully considered it. She planted a seed that started growing in my mind on the spot.

Having already completed two transactions with me, she no doubt had the real estate wheels churning in her head. And ultimately we'd do two more deals. But first and foremost I think she honestly felt bad for me, and was more concerned about my state of mind than my wallet.

Ray of hope

In my gloomy state during the bleak and depressing winter of 2004-2005 it seemed to me that everyone at every dinner or social gathering I attended was thinking, "Poor Greg...he wanted to go into space so badly and now he can't." Exactly the kind of pity that I hate and don't want.

Maybe it was all in my mind, and they were thinking nothing of the kind. No doubt they wished me only the best – assuming they thought about my plight at all!

Just in case things turned around I started taking Russian lessons from a kind woman named Alla Lapacheva to keep up with my language skills. She did what she could to help me, but she could see how depressed I was over being excluded from the space program.

Limited aptitude and a negative attitude are a pretty bad combination in any field of study. I certainly wasn't among her best pupils!

Over at Sensors II, with Marshall and Chris running the show, business and revenues were picking up. But I wasn't contributing much at this stage, so even that good news didn't have much effect on my depression.

One of the darkest moments of this whole period was in February, when Eric Anderson met with me for further discussions on my spaceflight contract. I had already filed the insurance claim to get my money back, but we both knew I didn't want a refund – I wanted to fly.

He wasn't very encouraging about my prospects. Discussions with the Russians weren't going well. It wasn't for lack of trying – Space Adventures had been working on the GMK for months, trying to get them to reconsider their decision.

Eric and his people were arguing that the spot on my lungs, which had caused all the fuss in the first place, had disappeared five months earlier, and had not returned. They were submitting documentation, X-rays, everything. The Russians were being stubborn, and progress seemed stalled.

His counsel to me was, "Just be patient." As a guy who could go head-to-head with me in the determination department, he should have known that this was exactly the wrong thing to say. One of the few virtues I neither pretend to nor aspire to is patience!

Signs of a thaw

As we got closer to spring, however, things started looking just a touch more encouraging. It was like when you're freezing to death in the snowy woods, and some sunlight peeking through the clouds gives a gentle hint of better days to come.

Eric began making veiled suggestions that the Russians might reconsider my medical situation. Susan James had begun showing me apartments in the Trump International. Sensors was on a profitable track. Even the 2003 Pinotage from my South African winery was tasting pretty good!

For some reason, throughout my life good and bad fortune both seem to come in waves. When one thing goes well, lots of things go well. When something important turns sour, like the medical discharge in June 2004, everything else seems to go bad at the same time. Was the cycle starting to turn up?

Acceptable lungs

Then the Russian medical committee invited me to a hearing on my status, arranged for me by Space Adventures. The hearing was on April 1 – April Fool Day, ominously enough. I asked Dr. Murphy to come with me as an expert witness, to help convince the Russians that I was fit for space flight, and off we flew to Moscow.

With Dr. David Murphy in Moscow, on our way to the Russian Space Agency doctors to plead my case.

There we sat with our piles of X-rays and medical tests outside the meeting room, anxiously waiting to testify before the committee. Just as we were about to go in and have our day in court, the door opened and someone said, "Dr. Olsen, we have decided to accept your lungs into the program." I'm serious – those were their exact words.

My first impulse was to say, "Wait a minute – I brought this well-known expert all the way from New Jersey to testify on my behalf. Don't you want to hear what he has to say?"

Then I recalled Bob Struthers' famous saying: "When you see the guy reaching for his wallet, shut up, stop selling and write the order!" So I shut up, accepted my victory, and left with Dr. Murphy and all our un-examined data.

I mean, who knows…maybe having Murphy sitting out there influenced the Russian doctors into admitting me. Whatever the case, every-thing worked out! (Like I've said a couple of times already…*don't give up!*)

Celebration time

Accepting my lungs into the program wasn't the same as approving the rest of me, but it was a start. Maybe the Russians would consider re-admitting me to the program after all. Things were looking up.

There was positive movement on the Sensors front too. It was about the same time as the Moscow meeting that we began discussions with M&A firms about possibly selling the company. Interest seemed high.

After nine months of living in a depressed state of mind, I could ac-tually smile again. As my 60th birthday approached, I was starting to feel optimistic, which is my normal outlook on life anyway.

My birthday was April 20. Ten years before my daughter Krista, along with several good friends, had given me a nice surprise party for my 50th. I was not expecting anything similar for the 60th. When she invited me to a small gourmet dinner with just a couple of close friends and family at her house, I was looking forward to a quiet, intimate celebration.

I hired a car to take me to the house because I wasn't sure whether or not I would stay over, and I was looking forward to sampling much fine wine. As we approached her driveway, I caught sight of some very familiar faces: my old advisors Bill Jesser and Doris Wilsdorf from UVA! Clearly, this was not going to be a quiet little celebration.

And for sure, it wasn't. Krista had arranged for nearly every signifi-cant person in my life to attend. George Antypas and Ron Moon from

Celebrating my 60th birthday with my daughters Kimberly and Krista.

California. Jerry Berman from Massachusetts. Most of my close friends and co-workers from the Princeton area.

To top it off, my daughter Kim had come all the way from Argentina with granddaughter Romina! What a fabulous surprise – you could not have done anything more to make me happy that night. To this day I am extremely grateful for that evening.

It's odd when you think about it. The worst nine months of my life began and ended with a birthday party, Jim McLaughlin's in July 2004 and mine in April 2005. What a difference between the first and the second!

Full acceptance

Although I was starting to get positive feelings about my chances of being readmitted to training, I didn't want to let my hopes get too high. I'd been shot down before. So I occupied myself with other activities while waiting for some news.

One of those activities was a joint presentation on the Sensors Unlimited infrared camera that Marshall and I had agreed to give at a conference in Scotland in early May. Eric had asked me to be on the alert for any news from Russia, but I kept my optimism in check.

I packed my usual carry-on bag and headed off for a 4-day visit to Scotland – with maybe a little golf mixed in. It was the type of business trip I've made hundreds of times.

It was on either Wednesday or Thursday, after we'd given our presentation, that I got an e-mail from Eric saying "GMK wants you to come to Moscow on Monday for a medical evaluation. I'd get right over there…."

Well, he didn't have to tell me twice: I was on the phone arranging flights from Glasgow to Moscow almost before I finished reading his message.

By Monday morning bright and early I was at the medical offices in Star City, ready to go through the usual thorough Russian exam. To my surprise all they did was take cursory pulse and blood pressure measurements, and then lead me into a meeting with about ten other doctors.

There seemed to be a more relaxed attitude in the room. It was not at all the dark and somber mood that I had grown used to.

The head doctor looked up and delivered a short speech on how seriously they take medical issues, how careful they must be with any cosmonaut candidates, and how carefully they had weighed my many deficiencies. After much deliberation, he said, they had decided to allow me back into the program, on the condition that no new medical problems arose.

What? Did I hear him right? Are they actually going to let me back in?

I was almost dizzy with delight! All those months of heartache and depression were over, all my work was not in vain after all. I was so elated, I could hardly breathe. I wasn't even upset by the head doctor's remark that "You should take it easy…after all, you're an old man!"

Old man, hell. I could swim, run and dance all night, even at 60. I was probably in better shape than this 50-ish doctor. But I wasn't going to let a remark like that spoil this fabulous day.

Thinking back, they must have made this decision days before. The meeting was a mere formality. I was immediately on the cell phone to Eric, asking him when I might start training. He replied that they'd like me to start ASAP – like *now*!

"NOW…?" I started to say. "Why, I don't have enough clothes, or toiletries, or my files, books…." Then the Struthers saying about shutting up when the guy is reaching for his wallet kicked in again. Hey, that stuff could all be shipped over. Why waste a week to go back and pack it all up?

Besides, who knew what might happen if I left. Maybe they'd change their minds again. I was having none of that. I was already at Star City and that's where I intended to stay.

And stay I did, all the way to my launch date, with the exception of a single week's training in August at the Johnson Space Center in Houston.

Back in Star City

Star City can be intimidating to first-time visitors. It's an official Russian Air Force base, and security is very tight. You enter through a checkpoint manned by armed guards. There's no way to zoom past them on the way in or out because of the maze of concrete barriers at the gate.

For me, though, the experience was different. I felt more at home. I'd been there already, and knew the instructors and some of the faces at the café.

Even better, I was staying at the "Profi" – short for profilactorium. This building served as Star City headquarters for NASA and the European Space Agency. It included living quarters for visiting cosmonauts. This was a more central location than my previous apartment building. There were lots of people to interact with, and it was closer to classes and near the NASA cottages.

Right from the start I met people involved in the space program, American as well as Russian. Many of the instructors recognized me, and seemed pleased that I had been able to recover from my setback. The Russians admire determination. They are certainly used to facing setbacks!

Another great experience was getting acquainted with Bill McArthur – Billy Mac, as he's universally known – the NASA astronaut on my upcoming flight to the space station.

I had met Bill briefly during my 2004 stint in Star City. One day I was walking past the Soyuz simulator when somebody inside shouted out, "Hey…aren't you Greg Olsen?" It was Bill. He had heard about the new space tourist and wanted to meet me.

Now, having been readmitted to the program a year later, I was walking to one of my first classes when I saw a guy on a bike wearing the same brand of jeans as mine – the ones that post your size where everyone can see it. These said 34x32 – also the same as mine. Just a coincidence, true, but when I looked closer, I realized it was Bill.

So I reintroduced myself. He remembered me and, after welcoming me back, offered to help in any way he could.

That kind of generosity was typical of Bill. But he also had a special interest in me as a passenger on the Soyuz because he had been the NASA Director of Operations (DOR) when Dennis Tito made his pioneering flight.

Back in 2000 Bill had been caught up in all the political flak surrounding NASA's opposition to civilian flights in general, and to Dennis Tito's flight with the Russians in particular. But he's fundamentally an honest and decent person, as everyone who knows him will attest. I think he respected what Tito was trying to do, and probably took some heat for it.

I thanked him for the offer of help, and gratefully accepted his invitation to come over to meet the guys in the famous "Shep's Bar" in cottage 3 of the American-style townhouses where the astronauts and NASA personnel lived.

Fitting in

It turned out that Bill and I had a lot in common beyond our jeans size. We both had two daughters, engineering backgrounds, and a love of good red wine.

After some beers with the boys, I got invited to join in one of their "pot luck" suppers, where everybody would bring something: salad, dessert, maybe some meat. When I volunteered to bring some of my wine, a hearty cheer went up.

The only wine they'd been able to get was the sweet Georgian variety that the local stores carried. But I still had my connection with the South African embassy, and could get as much of my wine farm's product into the country as I wanted. With my value to the community firmly established, I was ensured of future invitations.

Bill enjoyed the wine, but complained about the tiny wine glasses that NASA had provided. As soon as I got back to my apartment, I sent an e-mail to my assistant Jennifer asking for a rush delivery of four full-sized Riedel cabernet glasses. I'm sure their arrival a week later, when Bill returned from a visit to Houston, helped seal the deal with him!

In addition to our taste for wine, Bill and I shared a love of Starbucks coffee. We weren't alone. What passed for coffee in Star City was too bland and weak for the NASA crew. If you went back to Houston for any reason, you had to bring at least a bag of Starbucks with you when you returned!

Bill showed me the ropes around Star City and in Moscow, like the concept of a "walk-around" beer. It is perfectly acceptable to drink beer in public in Russia. You stick one in your pocket when you walk around the streets of Moscow. And no self-respecting Russian man – or NASA astronaut for that matter – would ever use an opener to pop the top. It just wasn't considered manly. Your thumbnail, a fork, anything but an opener!

Bill was always there when I needed a friend. Through him I got to know many of the astronauts, plus the Russian and NASA personnel.

None of this was required or expected. He did it because he's Billy Mac. He saw someone who felt a little out of place and needed a little orientation and support, and took it upon himself to provide the necessary guidance. Bill is right up there on my list of lifetime mentors.

Training regimen

Bill also went the extra mile in helping me with pre-flight training. For example, he shared non-confidential NASA learning aids that helped me navigate the sometimes obtuse Russian training manuals.

Here, I'm sure, he had another motivating factor beyond being a good guy. I was supposed to launch with him in October. It was clearly in his interest to have a well-prepared "space participant" flying beside him.

By August I was training with Bill and mission commander Valeri Tokarev as a crew of three. If I screwed up in a training drill, Bill never yelled or criticized. Instead, he'd say something like, "Hey, don't worry, Greg …most people mess that up first time. Try doing it this way instead."

That's a real lesson in leadership. I now try to incorporate this approach in my own dealings with people, though I probably don't do it as well as the master I learned it from.

A typical day of training went something like this:
- Up at 6 a.m.
- Run two miles around the beautiful Star City Lake
- Breakfast in the cosmonaut cafeteria where I had my usual "zapekanka" – something like a cheesecake – a boiled egg and some tea
- Classroom sessions 9-4 with a one hour lunch break
- Physical training (weights and swimming) 4-6
- Dinner 6-7
- Homework and bed

Classroom topics included a multitude of safety and emergency procedures, such as what to do during fires or leaks of toxic gas or cabin pressure; how to use the radio and intercom; Ham Radio procedures; photography and video equipment; electrical systems training; the design of the rocket and its operation; and orbital parameters and procedures for achieving orbit.

They also gave me very basic instructions on how to fly the Soyuz, although I was not expected – or allowed – to actually operate the rocket unless there was some truly dire emergency.

And, of course, there were the Russian classes. These ran at least two hours a day, and sometimes four. My frustrations with learning the language persisted. I really wanted to become fluent, but I just don't have what it takes to do that, in Russian or any other foreign tongue. However, I will say that the teacher, Svetlana Rosanova, went out of her way to help, including using photos of my grandkids to teach me the Russian words for family members.

While the classes were intensive, the exams were positively intimidating. You took most of your tests orally, standing in front of a panel of three unsmiling Russian Air Force officers who lobbed the questions at you.

They took their job seriously, as well they should. Your fellow crewmembers' lives could depend on you knowing your stuff, not to mention the success of your mission.

Local justice

You've got to get some relief from the pressure when you're training as intensely as we were. Otherwise you risk losing your edge. During my two tours of duty at Star City, going to Moscow for the occasional weekend gave me a good way to blow off steam.

It also provided some insight into the Russian way of doing things. As I mentioned earlier, the Russians are high-risk drivers. I got the distinct feeling I would be safer in Soyuz than I was on the highway. Everyone seems to do 100 miles an hour on roads that really aren't up to that kind of speed.

As a result, I got to see how the traffic cops operated in Russia. In fact, I witnessed them in action at close range, because we got stopped four times while I was there. No, I wasn't driving. I'm not that crazy.

One stop that stands out is when we went through a red light and got pulled over by a police cruiser. The guy who was driving the car chatted with the cop for a while. Unfortunately, my Russian wasn't good enough to follow the conversation.

Then the driver reached his right hand into his pocket, pulled something out, and used it to shake hands with the officer. After that we simply drove on our way again. You didn't need to know the language to understand what had just happened.

We got off relatively light in terms of the money that was exchanged. If the driver had been drinking the "instant fine" would have been a lot higher. People have told me that they always carry a thousand-ruble note with them in case they get stopped for DUI.

In their defense, the police were hardly earning a living wage at the time. Collecting a little extra in a traffic stop helped them supplement their income. Their supervisors might warn them against accepting bribes, but everyone knew it was happening.

In fact, the system was so ingrained that transgressors who failed to compensate the police for their trouble not only didn't get off easy, they invited additional charges. We once drove past a little roadside drama. Two cop cars had stopped a big sedan, and they were giving the car a real going-over. They were using mirrors to look at the undercarriage, shining flashlights in the gas tank, and so on.

"The driver didn't offer them any money, so they're looking for other violations they can add to the charges, such as defective equipment," my driver Misha said. "They'll find something, too, even on a Mercedes."

I've been back to Russia several times since, and these payoffs don't seem to be as prevalent any more. As a people the Russians seem to value order, and I'm guessing they've found ways to rein in the extracurricular collection activities, perhaps with better salaries for their police.

Reaching for the stars

Finally the day I'd worked toward for over two years was at hand. It was a Sunday morning in late September, ten days before the launch, and we were leaving Moscow for Kazakhstan.

The Muscovites threw us a party that morning, complete with food, vodka and cognac, and an official read a proclamation in our honor.

Russians follow their space program with much more enthusiasm than we do in the United States – almost to the point of fanaticism. They also revere their cosmonauts. Even I had become something of a celebrity – I signed lots of autographs.

Hundreds of well-wishers lined the streets along our route, many bearing flowers and all of them waving to Valeri, Bill and me. It was amazing – they certainly celebrate life in Russia differently than we do in the United States.

The same enthusiasm was in evidence at the launch site, where thousands of people always turn out for the twice-a-year launches. My participation made the crowd a little bigger – some 30 friends and family members traveled to Moscow to see me off, and most continued on to Kazakhstan for the launch.

I knew that they were happy for me and proud of what I'd accomplished. But in the pre-launch days one of the group, my oldest sister Amy, was very much on my mind. I was worried about her. On the flight from the United States to Moscow she had experienced a violent coughing spell that cracked three of her ribs.

Amy had followed in dad's footsteps as a heavy smoker. She had major respiratory problems and had already suffered two collapsed lungs. I knew that the chances of her having a medical emergency en route to Kazakhstan were high, yet she kept insisting that she was going.

As much as I love her and value all that she did for me as a kid (sometimes filling in as surrogate mother), I did not want her risking the trip to Kazakhstan and possibly ruining the occasion for herself and everyone else.

If I felt any pre-flight stress at all, it was because Amy was adamant that she wanted to be there for her brother. Instead of focusing on the flight, I found myself preoccupied with keeping her off the charter jet that was taking my extended family from Moscow to the launch site.

Fortunately, she finally agreed to stay behind in Moscow, after a lot of intervention by me and other family members.

Launch preparations

Finally, my wildest dream was about to come true, in full sight of cherished friends and family.

Of course they arrived later, a day before the launch, too late for any direct contact with me before we took off. It's standard procedure for the crew on a Russian space flight to be quarantined for a week before launch. The last thing you want to do is pick up a cold from a well-meaning relative and then carry the virus up to the space station

My grandson Justin, nearly four years old at the time, couldn't un-

In quarantine, just before launch, with grandson Justin.

derstand that at all. When my 30 family and friends pulled into Baikonur on their bus, the Russians allowed me to go out and greet them, but warned me not to touch anyone.

So out I went to the bus, warning everyone inside that I couldn't have direct contact with them. All of a sudden Justin bolted off the bus and came running to give me a big hug, as he normally would. I had no choice but to run away, with Justin in hot pursuit.

He was really shocked. "Grandpa! How come you are running away from me?" His mom came to the rescue, scooping him up just before he reached me, explaining what quarantine meant.

Later that evening we held a press conference and answered people's questions from behind a glass partition, using microphones. Justin was in the audience. He came up to greet me, and placed his hand on the glass. I immediately placed mine on the other side to give him a "high five" through the window. A Russian reporter caught the moment in a fantastic photo that made one of the Moscow newspapers.

On my return from the space mission someone gave me a copy of picture. I was so taken with it that I contacted the reporter and asked if I could get a copy. Talk about a nice guy – he sent me printed copies plus an electronic file! It's a photo I cherish and keep on display in my office

My final contact with Justin before the takeoff came as our bus to the launch pad was about to emerge from the crowd and head to the waiting Soyuz rocket. As we got to the edge of the crowd, I heard a loud knock on the window. It was Justin, sitting on his father's shoulders, banging on the window to wish me well. Having my little buddy there to share the moment when I fulfilled my dream was really the icing on the cake.

Toward the stars

We arrived at the Soyuz TMA-7 rocket before daybreak on October 1, 2005. The three of us climbed up the 24-foot ladder, turned around, and waved to the two thousand spectators in the crowd below.

Decked out in my 22-pound spacesuit with helmet and rubberized gloves, I scrunched my 6' 1" body into position before I was strapped into the "third seat" of the space capsule, on the right-hand side. In repose the seat tilted back so the intense gravitational force of lift-off would be concentrated on a cosmonaut's stomach, rather than the head. It was pre-flight time, the last stage before beginning the ride of a lifetime.

Cosmonaut Valeri Tokarev occupied the flight commander's seat in the middle, to my immediate left. On the far left sat Bill McArthur, four-time NASA astronaut and my unofficial space mentor. As the only civilian, I was honored to be in such illustrious company when I joined the elite club of space travelers.

Pre-launch checks

Locked in the seven-foot-wide capsule, the three of us tested our space suits for air leaks by hooking up to the air supply and measuring the time it took to reach a certain pressure. When everything checked out OK, we were given the go-ahead to start flight prep.

It took over two hours for Valeri and Bill to go carefully through the pre-launch checklist in their notebooks. I followed along in my manual, also written in Russian, with my marginal jottings in English.

My job was to be nearly invisible. This went totally against type for a guy who thinks of himself as an active participant in life, but I fully understood and accepted my defined role. My traveling companions knew that I was available to help out but I kept quiet while they went about their work. They had their hands full; this was no time for small talk or stupid questions.

Occasionally I held Valeri's pointer for him, when he wasn't using it to activate switches high up on the control console. Sometimes I held his notebook so he could have both hands free for the intense systems checks he was performing. I was thrilled to do any menial task that helped, even if it was just keeping an eye on the cabin pressure and oxygen readings.

For once in my life, there was nothing I absolutely had to do, no tasks I had to perform. I could give myself over to the total enjoyment of having accomplished a major life goal. I've never felt more peaceful or more tranquil in my entire life.

If I worried about anything, it wasn't danger or even death. It was the possibility of having the launch called off at the very last second. So little had gone smoothly during the 18 months prior to arriving at the launching pad that cool autumn morning that I had a nagging conviction that I couldn't be sure of blasting off until the rockets fired. Suppose the Russians discovered some obscure medical irregularity in a urine test, or quibbled over the interpretation of a preflight EKG?

Call me overcautious, but I had instructed Jim McLaughlin (who was there to witness the launch) not to authorize the final payment for my trip into space until the rocket had cleared the launch pad. Once I got up in the air, I knew for certain that there was no calling me back.

Liftoff

Finally the preflight was done. Those two cups of tea I'd had earlier were making their presence known. Good thing we were wearing adult-size diapers. It would be hours before we could visit the primitive bathroom in the habitat section of the capsule.

Then the engines began to fire, and the launch was under way. It was on time, pretty routine for the Russians, and we rose into the sky above the Baikonur Cosmodrome.

Because the Soyuz is symmetrical, there was little vibration on takeoff, and the ascent was smooth. Thanks to the four layers protecting us from the noise of the liftoff explosion, it was also surprisingly quiet, especially compared to the deafening roar I'd experienced as a spectator at the launch all those months ago.

Several minutes after takeoff, 50 miles above the earth, I heard a loud boom and felt the jolt as the exploding bolts holding the shroud to the spacecraft fired and blew the covering away. Suddenly I could see out the window to my right, and witnessed the huge sphere of the earth receding. What a view!

The rocket continued its ascent. I tried to lift my left arm and experienced the force of acceleration making it feel as if it were three times as heavy as it is on earth. I have a broad background in physics, and I've even taught elementary orbital mechanics, but understanding the effect doesn't prepare you for the experience of feeling it in real life.

Once we reached orbital velocity, it would be like reaching the top of a roller coaster, when you feel a brief moment of weightlessness before going down – only our weightlessness would continue.

But well before the rocket reached orbit, I experienced a different, non-physical kind of weightlessness. It was the psychic lightness of having a huge burden lifted from my shoulders. Mission accomplished! I felt joy, exhilaration, happiness.

The next ten days belonged to me. It literally didn't matter what else happened. I had made it. I had reached the goal that had obsessively consumed me for the previous year and a half – and done something I had dreamed about since I was a youngster. I was actually going into space. Nobody could take this away from me now.

As we flew higher, we monitored our approach to a state of physical weightlessness with the lowest-tech instrument employed on the spacecraft. It was a three-inch plastic "troll" doll.

This is a Soyuz tradition. Before every flight, a Soyuz commander chooses a mascot. About ten minutes after takeoff the doll begins to float by itself in the capsule, as if floating in water. When our troll started drifting around, we knew that we were officially in orbit.

Bill and Valeri still had a lot more checks to do. As a safety precaution we kept our suits on for a few of the 90-minute orbits of the earth. Occasionally Valeri would turn to me and ask how I was doing. I would say, "ya chust-vay-u hor-a-show," which meant "I feel good." And I did.

In the quiet time, I thought about my two daughters and four grandkids, including the twins my pregnant daughter Kim was expecting. I had brought along personalized dog tags for each of them with their names and the date of the mission, all signed "flown in space with love" by Dad or Grandpa. I planned to give the mementos to them to them on my return.

When the little ones grow up they will learn what their granddad did. I hope they will appreciate my unusual achievement. Certainly my daughters knew what it took for me to get where I was, circling some 220 miles above the earth. I had 10 days to contemplate it myself.

Floating free in the ISS.

Holding a flag from my South Africa vineyard. It now hangs there.

Soyuz vehicle docked at the ISS.

9 Re-entry

A truly amazing adventure

Greg Olsen has become the third civilian in history to travel in space. He...trained for two years...and in just 10 days traveled over three million miles in space and completed over 100 orbits of the earth.

—*Good Morning America (ABC-TV)*, 10/17/2005

There we were in orbit, hurtling along at an incredible 17,500 miles per hour, 220 miles high. It had taken less than ten minutes to get there. But even though we were in orbit, we had to remain in our spacesuits, strapped into our seats in order to maintain our emergency oxygen lifelines.

Below us was the beautiful planet we call home. From that height the first thing that struck me about Earth is how blue it is. But then the scientist in me took over, and I realized that it's the atmosphere that's blue, not the planet. I was looking down at the sky!

I vividly recall looking out the window and suddenly seeing this massive cloud formation, with an unmistakable funnel inside: a hurricane. I don't know why I didn't photograph it with my small digital camera. I stared in awe instead.

I was also amazed at how small Earth is, and how fragile. Our atmosphere is barely 10 miles thick, and it's the only thing between us and the endless void. I wish everyone could see this for themselves. They'd immediately grasp the need to protect this crucial, miraculous membrane that supports life on the planet.

Just being there was the best experience of my life (after the birth of my daughters Krista and Kim). If I could do it again, I would go back in an instant.

This trip was more than a passing fancy. It spoke to some deep human need to explore, to advance, to achieve. When I talk to kids about my adventure in space (more about that later), I tell them that you have to pursue your dream. You can't let yourself get discouraged, or settle for things as they are. If you set your mind to it, there's little you can't accomplish.

Space was that kind of experience for me. However small a part I played, I was fortunate to be able to participate in the early stages of mankind's next bold adventure, the exploration of space. It was a great privilege for me.

Next stop: the International Space Station (ISS).

Around the world in 90 minutes

All told I spent 10 days in low earth orbit – eight of them on the ISS. You certainly can't top the view, from the awe-inspiring globe of the Earth to the stars and planets visible against the black vacuum of space. But my most lasting impressions all revolve around the sensation of weightlessness.

When we first achieved orbit we didn't have time to consider our new weightless freedom. I was holding various procedure manuals while Valeri and Bill performed the numerous chores necessary to establish orbital flight, such as a detailed leak-check of the vehicle.

Once everything was found to be "nominal" (normal), we could detach our oxygen lines, open our visors and gloves, detach our seatbelts and float!

I knew right away I was going to like it. There had been a slight sense of flotation even while we were strapped in. Now we were simply drifting at will. That doesn't mean we moved without meaning to – we might not have had any weight, but we still had mass. It took a little effort to begin movement, and also to stop that movement.

We were experiencing weightlessness. That's the accurate term for this condition, not "zero gravity." True zero gravity occurs only very far away from any massive body, or at a neutral point between two masses, where their attractive forces just balance. One of these "LaGrange points" exists nine-tenths of the way from the earth to the moon.

We weren't even close to being that near the moon. The Soyuz and the ISS were orbiting the earth only 220 miles high. At that distance the pull of earth's gravity was essentially the same as it would be on the earth's surface.

Yet there we were, floating around. At 17,500 miles per hour, the earth's gravitational force is balanced by the outward "centripetal" force of the orbital motion. This not only keeps you in orbit, it makes you weigh essentially nothing. Everything – you, your seat, the vehicle – is in the same orbit, so there's no sense of weight, in spite of the gravity. The so-called "zero-g" flights for astronauts use the same principle.

Needless to say, these scientific facts were hardly at the top of my mind as I unstrapped. I was simply enjoying the delightful experience. It's hard to describe what it's like to be unconstrained by gravity. Floating up into the Soyuz spherical habitat module, about 7 feet in diameter, I felt free – truly free – WOW! After all the worries, waiting, and training, my dream had finally come true.

I was floating for a full 10 days. There was no up, or down, or left, or right. Every direction was equivalent. It was a blast!

When we got to the ISS I was in an area about the size of four tractor-trailers, and I could float its entire length just by pushing gently off the back wall. In floating down the station I'd rotate 90°, transforming the ceiling into the right wall, the right wall into the floor, and so forth.

Even while I was getting oriented, which took four or five days, the ability to move easily in every direction was truly liberating. I loved zooming up and down the station.

The absence of gravity not only expands your access to physical space, it changes the way you look at movement. You leave your ordinary two-dimensional existence behind. Going "up" no longer takes effort. You operate freely in three-dimensional space.

There are some downsides, of course. Eating or drinking can be a challenge, and personal hygiene (including going to the bathroom) take very different procedures from those you use on earth.

Another of the side effects of weightlessness is that blood tends to accumulate in your head. As a result, your body decides there is too much blood in your system and shunts it to the kidneys for filtration and disposal as urine. When that begins to happen, you have to pee more often! At times we were lined up at the urinal hose (really, a vacuum cleaner hose where the suction pulls in the liquid). You lose about 15% of your blood supply in space that way.

But on the whole, gravity-free existence was a great experience. It turns even the most ordinary activity of life into its own kind of adventure. When you want to sleep, for example, you don't use a bed. You get into a sleeping bag, tether it to a wall or the ceiling, and literally float off to dreams. (If you don't tether yourself to something solid, there's no telling where you'll wake up.)

Or consider what happens when you move. If you start typing something into a computer, just by tapping on the keyboard you're exerting enough pressure to push you to the ceiling. Newton's third law of motion – for every action there is an equal and opposite reaction – takes full effect under conditions of weightlessness!

A standard cosmonaut trick is to push off from one end of the vehicle and make it all the way to the other end without touching the sides. I never quite made it – it takes a real expert like Sergey Krikalev, our commander on the return flight, to pull that one off!

Another striking feature of weightlessness is the way physical objects disappear, seemingly of their own accord. I must have taken a hundred or so pictures with my pocket-sized digital camera to capture and preserve my experiences. I kept the camera – where else? – in my shirt pocket. One day it wasn't there. It had floated out of the pocket and drifted away.

I looked for it before I left the ISS, but with no success. When I got back to earth I sent a message to my crewmate Bill McArthur asking him to keep an eye out for it. Of course it really wasn't the camera that I wanted, but the memories captured in all those pictures. When he found it, I asked him to download the photos, since he wouldn't be back for a several months.

Having your pictures sent to you from an orbiting space station was a really cool idea, but while it was easy to do, bureaucracy found a way to complicate the process. Bill was working for NASA and I was a private citizen. The photos were private property, not NASA material. There were issues revolving around them having to bill me for providing the service.

So we compromised. I agreed to give NASA ownership of my photos. Since all NASA photos are in the public domain, freely available to anyone, they could send me the pictures I took without having to bill me for the privilege.

Yeah, that's right … the only way to get my photos back was to agree they weren't mine. Bureaucracy is wonderful that way. It's nice, though, that those pictures are now available for everyone to see.

Orbit and return

Crew members are tightly scheduled during their stay on the ISS. Mission control in Houston or Moscow sends up a radiogram each morning that details every hour for every person, including their assigned free time.

The day begins with breakfast around 8, usually combined with a status chat with the ground. Then crew members perform their scheduled duties, be it changing a filter, performing a science experiment, taking photographs, or the daily medical conference with a flight surgeon in case they have any "ailments." This is done in private so that crew members can talk candidly about their situation.

Although I was flying with the Russians, they graciously allowed me to use Dr Richard Jennings as my flight surgeon. He knew me and my medical situation better than anyone.

We all got together at mealtimes. While breakfast was a time to reconnect with the rest of the crew, do "com checks" with the ground, and review what was in store for us the rest of the day, lunch was more of a casual affair.

But dinner was our social time. It was more relaxed – our work was usually completed, we were a little tired, and we were more interested in enjoying our "camp" food and getting to know each other a little better.

I recall one night we were all discussing what food we were looking forward to enjoying when we returned to earth. John Phillips wanted beer and pizza. Sergey wanted real coffee, so he could smell the aroma of a fresh brew, instead of just swallowing instant.

Personally I would have traded the chance to stay up in space longer for any food on earth, but I chimed in with steak and red wine. It was fun, and a good bonding experience, just getting to know each others' thoughts and wishes.

Then it was back to our own pursuits before bed. Lights out occurred at 11 p.m., although some of the crew continued to read documents or spend time on the computer after that.

I should mention here that we kept track of time by setting the clock on the wall to Greenwich Mean Time. Since we were orbiting the earth every 90 minutes, there was no way to use sunrise and sunset as a reference.

Sleeping felt wonderful. It was…just a floating feeling, with my arms held out at my sides. To get in the mood I used to listen to opera, primarily Wagner's Ring Cycle series, on my $15,000 iPod. (That's how much it cost to flight certify it!) The strains of "Ride of the Valkyries" would remind me of the thrill of our Soyuz launch. Occasionally I would fire up some Chuck Berry or George Thorogood, two of my rock 'n' roll favorites. Hank Williams, Sr would round it out with some classic country.

The first two nights it was difficult to get to sleep, due to the excitement of being there. But after that it didn't take long to fall into dreamland. Occasionally rotations of the solar panels would wake me momentarily with a creaking noise. It reminded me of the sound of a ship's ropes.

Waking up was just as enjoyable as going to sleep. You open your eyes and realize that you are floating. I usually arose around 6 a.m., before anyone else, and went right over to the window. I would just stare out at Earth and, aided by a world map, try to figure exactly where we were. Occasionally I'd take photographs. That quiet time, all by myself, was extremely relaxing and peaceful. Sometimes being alone can be its own reward.

I did have a lot of free time on board, but I was never bored. I mean, just looking out the window – trying to figure what part of Earth you are passing over – who could get tired of that?

My time in space went by fast. I kept busy with the medical and scientific experiments I'd agreed to perform for the European Space Agency. I also had fun communicating with the students at Princeton University, and with the high school kids at Ridgefield Park High in New Jersey and Fort Hamilton High in Brooklyn.

During those 10 days, without ever really being conscious of speed, I traveled a nearly unbelievable distance, at least when viewed from the perspective of Earth. Soyuz and the ISS circled our planet over 150 times, making a complete orbit approximately every 90 minutes. The orbits covered a total of over 3,000,000 miles.

And then it was time to leave. Bill and Valeri were on the ISS to relieve an astronaut and a cosmonaut who had completed their tour of duty, so I was going back to Earth with cosmonaut Sergey Krikalev, who had already spent over six months in the weightlessness of the ISS. During my stay he set the world record for total days in space: 803 days. NASA astronaut John Phillips was the other returnee.

As commander of the return mission, Krikalev had to deal with a lot of anxiety the night before our return to earth. The Soyuz must be packed very carefully since the center of mass actually affects where the capsule lands.

We also had very little room to spare. During launch the vehicle has three sections, so there's some extra space. But the descending Soyuz is missing two thirds of that volume, and space is at a premium. Krikalev carefully balanced the vehicle, with much communication from the ground. All I could do was watch.

One issue arose around me and my medicines. The Russians had insisted that I take "Spiriva," which is an inhaler to aid lung capacity for people with chronic obstructive pulmonary disorder (COPD). This was packed in a container about the size of two books – not a lot of volume on earth, but a considerable size on Soyuz.

Ground control was insisting that it be included in the descending payload and the commander was having trouble finding room for it. We're talking about the kind of space you find under a car seat, in an economy car! I kept telling him that it really wasn't necessary but the people on the ground kept insisting that it was. I was embarrassed that so much time was being spent on my behalf on such an insignificant issue.

Finally, I suggested that I keep enough in my spacesuit for the descent and landing (it was only three hours for Chrissake! The inhaler was once a day). They agreed – problem solved. I think they were worried that if we had an emergency landing and were stranded for awhile, I would need my medicine. I'm glad that reason prevailed here. In fact, it was no big deal even if I didn't have that inhaler for several days.

Ironically, according to my pulmonary specialist Dr. Murphy, my biggest issue wasn't my COPD. It was my bullae – blister-like protrusions on the lungs that can expand and rupture if there's a rapid decompression. Neither the pleurodesis I'd had nor Spiriva protect you from this problem.

I don't think Krikalev got much sleep before we left. None of us did. When we got ready to undock, we had trouble balancing the pressure between two modules, which could indicate a potential leak. We checked and rechecked everything, and finally the ground decided that the pressure difference was small enough to ignore. Slowly, we pulled away from the ISS, first aided only by the springs in the docking mechanism, then by our thrusters, which fired once we got far enough away from the ISS.

We completed a little over one orbit and then began the descent process, jettisoning the instrument module below our seats and the habitat module above our heads. From time to time I would glance at the instrument panel to try to see our position, velocity, g-forces and cabin pressure.

I don't remember exactly when, but I began to notice the total air pressure in the capsule (normally 760 mm) was starting to drop. It went from 740 ... to 730 ... then 700 over several minutes. Commander Krikalev's eyes showed he was also picking up on this and I could see him checking this readout more frequently, in addition to the thousand other tasks he had to do. This was a slow decompression – exactly what had worried Dr. Murphy.

We had simulated this situation many times during training. I thought I knew what to do, and I was grateful to the Russian instructors who had drilled us on how to respond during training and reprimanded us if we didn't do it exactly right. When ordered to do so by Krikalev, I was to open the oxygen valve and keep it open until the total pressure got back near 760 or Krikalev commanded me to close it. It was down by my right foot, and I was the only one who could reach it easily.

So I kept my eye on that gauge and my ears open for Krikalev's command. When the pressure got below 650 I knew it wouldn't be long. At 630 mm, as best as I can remember, Krikalev shouted "Olsen – kislarod!" I opened the valve.

It wasn't easy. Spring tension in the valve, combined with the thick glove on my hand, made it difficult to hold open. But it was my duty and I was not going to fail, even if my arm fell off! I watched hopefully as the pressure began to increase. When it got somewhere above 730 Krikalev told me to close it. Thankful for the reprieve, I pushed the valve back in, and breathed another sigh of relief as the pressure seemed to hold steady.

However, opening the oxygen valve can create another problem: too high an oxygen content. If there is more than 40% oxygen in the atmosphere of the Soyuz capsule, it will automatically depressurize, and the emergency oxygen supply will feed into our spacesuits to pressurize them instead. (That's why you wear them – without pressure your blood will literally boil.)

This safety measure stems from the Apollo 1 disaster in 1967, when three astronauts died in a fire during training. They were using 100% oxygen, which is very flammable – if there's a spark or electrical short the situation turns deadly in a hurry. I nervously glanced at the partial pressure for oxygen: we were at about 34%. Enough of a safety margin not to have the emergency oxygen turned on in our spacesuits, assuming things stabilized.

Fortunately they did stabilize, and there were no further incidents during our descent and landing. Unlike the NASA Shuttle, the Soyuz spacecraft's re-entry capsule has little flight capability. Parachutes lower

L to R: Video feed from the ISS; after the landing; out of the capsule.

it through the atmosphere to a landing spot, so we had little control over our direction during descent. But we landed right on target, with a helicopter circling around us during our final descent.

A Russian recovery team was waiting to pick us up as the capsule's parachutes lowered us through that precious layer of atmosphere down onto the ground in Kazakhstan.

My great adventure was over.

Days later, investigators examining the pressure problem said that they had found a mark on a hatch seal. They believed that the pressure problem resulted from a strap that had become jammed in the hatch after we closed it. I remember looking at the hatches and not seeing anything there – that's also part of our training. But the investigators said we wouldn't have been able to see the jammed strap. I'm just grateful that all ended well, thanks to Sergey Krikalev's alertness and quick action.

Earth matters

Successfully completing my space flight was undoubtedly one of the highlights of my life. To make it even sweeter, the second sale of Sensors came through almost simultaneously, yielding another financial windfall for me. There wasn't much to complain about!

And yet I had this distinct feeling that a big chapter (maybe even "Part II") of my life was coming to a close. What lay ahead was uncertain. What do you do for an encore after you've gone up in space and then sold your company for the second time?

To be honest, I wasn't all that worried about it at the time. Nor was I particularly bothered by the question so many people asked me (and still do), which usually goes something like this: "How can you match that one?"

Well, how can I … and should I have to? Some really good things had happened as a result of my working hard all my life, and this was the absolute top experience. Why not sit back and enjoy the rewards – at least for a while?

I spent the latter half of October 2005 doing PR tours, parades in Star City, TV and press interviews for USA Today, CNN, CNBC, Good Morning America, and the like. And then it was over. I was back on the ground, acting like a normal earthling again. That was good, too.

I finally returned to the US from my completed space journey on Halloween 2005. The flight from Moscow landed at JFK and I went straight to my apartment in the Time-Warner building in New York. Susan James of Trump had invited me to a Halloween party, so to get ready I relaxed in the Jacuzzi whirlpool – one of the few times I ever used it.

Living spaces

It felt really good. There was a TV on the wall above the Jacuzzi, and a spectacular view of lower Manhattan through the window that was even better at twilight. Looking at that cityscape, sitting in a hot tub, enjoying a glass of wine – sometimes being by yourself is the best treat of all. This bathroom was one of the things that did come out the way I wanted – why not enjoy it?

All in all the view from the Time-Warner building was every bit as breathtaking as I had envisioned. If the decor had come out more to my liking, it would have been hard to give up that apartment. But my original feeling still prevailed: this apartment was not me ... I wanted a change.

Susan had actually put the place up for sale and in the meantime had sold me another space in Trump International while I was in training. I bought it sight unseen. Another mistake – I didn't like the Trump unit very much either. But this time there was no feeling of depression. We were able to resell the Trump unit for no loss (Manhattan real estate was still in good shape back then), and for the time being I was still in the Time Warner apartment.

It wasn't just the apartment's décor that bothered me. The whole building felt too big to live in. It was situated on top of a shopping mall (which didn't excite me). It took two separate elevators and about a 10-minute walk to access the pool.

Time Warner's staff were helpful and nice, but the way it was managed left something to be desired, in my opinion. There was no room service – no newspaper on demand – no valet parking either. My daughter had to park her car down the street and schlep three young kids in strollers to get to the building.

I longed to be across the street at the Trump, where I really liked the facilities and the people. It was a much more intimate, cozy building. As soon as my daughter or I pulled up in our cars, the doormen recognized us and were eager to help. They're hands-down the best staff in Manhattan.

Finally we found a buyer for my Time-Warner unit. The sale was completed in March 2006 and I was once again a man without a home in Manhattan. It felt weird going into the city and staying in hotels instead of at my own place.

Susan kept showing me units at the Trump, but after my initial abortive attempt to relocate there, I was going to hold out for exactly what I wanted. It finally appeared in May of that year.

It was huge: over 5,000 square feet with four bedrooms on one of the topmost floors. The living room alone was 50 x 30 feet – I could envision a dance being held there. The whole space occupied half of a floor, for a real wrap-around view of Central Park, lower Manhattan, and the Hudson River! Plus, it was sparsely furnished – actually decorated pretty much the way I would have done it. A few couches, a big TV and beds, no frou-frou stuff.

She had actually shown me this apartment earlier in the year, but I guess the first time I looked at it I was overwhelmed by its size, thinking I didn't need that much room. This time I really liked the feel of it, and remembered my old admonition: "When it feels right, DO IT! Don't second-guess yourself."

So I made an offer there and then, with a side deal for the sparse furnishings. The owners declined to sell the furniture, but did accept my bid on the apartment. As it fortuitously turned out, the price was almost dollar-for-dollar equal to what I got for the Time-Warner unit.

So two years later, after much aggravation, and with no money lost, I made my move back to the Trump … and to this day I've never regretted it. It's exactly what I wanted, and it offers by far the best view of the Macy's Thanksgiving Day parade I've ever seen.

Launching more dreams

Almost without my realizing it, life was moving into the next stage. I'd been an entrepreneur for over 20 years, and had nurtured two companies to success. I'd realized my crazy dream of going into space. Persistence had paid off.

When Goodrich bought Sensors Unlimited, they didn't even list me as a "key executive," subject to non-compete clauses and the like. But call me single-minded, or set in my ways, or just plain workaholic, I still had the urge to be involved in starting new technology businesses.

My plan was to get my affairs together and make a full-time pursuit of GHO Ventures. I haven't mentioned GHO Ventures before, but it had been around for a while. I started it as sort of an afterthought following a mid-2001 meeting with some guys who were starting up a company called Princeton Power Systems.

It happened like the Bruce Springsteen line: "I was walking out – he was walking in." One morning I was on my way into Starbucks in Princeton (so much of this story seems to revolve around Starbucks) at the same time as a guy with a vaguely familiar face was leaving the store. We had passed each other several times before. This time, the guy turned to me and said, "Hey, You're Greg Olsen, aren't you?"

"Yes," I replied, "and you're Ed Zschau, right?" We had seen each other at talks, conferences, and other events, but had never formally met. He stepped back into the shop and sat down to chat with me over more coffee.

Ed mentioned that he taught a course on entrepreneurship at Princeton University, and four of the students had just won his annual business plan contest with a really good business idea. Might I be interested in speaking with them? Sure, I replied.

It's this 24/7/365 attitude I have: you never know when a good opportunity will appear, so you always have to have your antennae up and ready. He agreed to bring his students to see me next morning.

We met, and I was impressed by their energy, enthusiasm, and smarts. They were all engineering students, all about to graduate, and all looking at very attractive job offers (the tech bubble hadn't burst yet). Yet all four were willing to work for modest salaries and no guarantees to commercialize a technology for inverters that one of their fathers had patented.

Inverters are electrical power devices that, among other uses, convert direct current (DC) electricity from a solar panel into alternating current (AC) like what you get from your household electrical socket. As an electrician's son I had a lifelong interest in most things electrical, so this immediately piqued my interest.

True, I was still CEO of Sensors, but Marshall was really running the company, and I was starting to think about what life after Sensors might be like. Would I start yet another company?

Hmmm.... I was in my late 50's and had done that twice. Watching others do a startup was starting to become more appealing than actually doing it myself. This could be my first "venture" into funding and mentoring others to do what I had loved doing for the last twenty years. Like many significant new directions in my life, this was neither planned nor the result of forethought – it just sort of happened!

After meeting with the students, Ed and I talked further, and we decided to jointly fund the new company, aptly named "Princeton Power Systems" (PPS). I like doing things this way, without a lot of agonizing about the decision or haggling over terms, and with a minimum of due diligence. Just find a good idea, and if everyone seems to like it, fund it!

As a side note, impatience with due diligence is not really a good trait. It's a fault of mine that has occasionally come back to bite me, as we'll see in the story of a startup we'll call ABC Company.

Eight years later, PPS has had its ups and downs, we might be in an energy investment "bubble," and it remains to be seen how successful they will be. But Princeton Power was the first of the many "angel" investments I've made since 2001. It launched me into my current role as CEO of GHO Ventures, which operates out of a small office in downtown Princeton, across the street from the university's Nassau Hall.

My initial plan was to use the company for making occasional investments. It also provided a bit of legal protection for two local properties that I had purchased as a real estate investment that year.

About every year I would add another "angel" investment to the GHO Ventures portfolio. By 2005 there were maybe six companies altogether, but I didn't spend all that much time on any of them.

I don't have an investment "strategy," and I *hate* reading business plans. I'm aghast at the thought of venture capitalists who not only read, but study over 100 business plans per year. Not me, buddy! I do not solicit business plans and generally dislike it when people refer others to me. I just seem to run across ideas and people who interest me, and when I do, I often invest.

GHO Ventures had grown to the point that it needed my attention. Fortunately, Jennifer Kennedy agreed to leave Sensors/Goodrich and work as my assistant in my new angel fund, and she's been here ever since. Her competence and organizational skills have played a big part in keeping my business affairs on track.

New possibilities

A cynic might say that GHO Ventures is Greg's way of trying to replicate the success of his earlier companies by spawning a bunch of new ones. Well, there's no sense denying that one goal of the enterprise is to make some money. It's not that I absolutely need more money, but I do want the business to be successful.

GHO Ventures also has a purpose beyond simply making money. It's a way of giving back. As I keep saying, I always worked hard, but I've also been very lucky. I've gotten financial support when I needed it most, and I've had fantastic mentors. With angel funding from GHO Ventures I can help other technology entrepreneurs get their businesses off the ground. And if they need a mentor to help them meet a challenge, I can give them advice or point them to someone better qualified.

There's potential for a bigger impact, too. It's exciting to think that GHO Ventures could be financing the next Microsoft or Google or other company that goes on to change the way we live while creating economic value for its employees and investors.

Promise and disappointment

As of this writing GHO Ventures has funded nine companies. They represent a variety of technologies.

- *Princeton Power Systems* (founded July 2001) produces electrical power conditioners to facilitate the adoption of clean power sources, and to allow precise control of industrial processes.
- *Eye Response* (February 2002) has developed techniques for identifying where people look and unique approaches to the best use of that information.
- *Achieve 3000* (July 2003) is the most powerful way to help students read, write and learn better.
- *Innovative Photonics* (August 2003) was founded to provide high-performance semiconductor diode light source systems based on proprietary spectrum slicing technologies.
- *Innovative Engineering* (April 2004) specializes in designing and manufacturing or assembling power delivery systems for industrial and medical markets.
- *Common Ground Recycling* (February 2004) produces a small, lightweight, and cost-effective tire-recycling machine.
- *ABC Company* (June 2005) – not its real name – was founded to

leverage its expertise in electricity storage applications to provide high quality distributed electrical power and storage systems.

- *Power Survey Corp* (March 2007) is a service company which uses reliable proprietary technology for the drive-by detection of hazardous stray voltage conditions.
- *E-Compliance* (October 2007) specializes in the delivery of innovative practical technology solutions to measure, manage and mitigate Health & Safety Risks in the quest for 'Zero Incidents'.

All of them were still in our portfolio at the start of 2009, though ABC Company was dissolved early in the year. We'll get to that in a minute.

I didn't start any of these companies myself, and I don't own any of them outright. My ownership ranges from 25-70%. There's no rhyme or reason to this. It just happens.

I hear about these companies from friends, business associates, other companies, venture fairs – just about anywhere! It's the same approach as you take to hiring good people. You don't limit it to 2 hours a day on alternate weeks – it just runs in the background 24/7.

Finding a winner

Take Power Survey, for example, the most successful of these companies to date. I found it almost by accident.

On Monday nights I often look through the magazines and documents that pile up on my desk while I enjoy some wine and pasta at Sotto, a favorite Italian restaurant in Princeton. One night in 2006 I was just about to sit down for a pasta-and-magazine session when Nancy Carnes spotted me and invited me to join her and her girlfriend for dinner.

Nancy is the wife of retired Sarnoff CEO Jim Carnes, the guy who tried to get me to run one of their spinoff companies just before I started Sensors in 1992. I was in the mood to be by myself that night, but not wanting to be rude, I sat down at their table. Besides, Nancy is delightful company. I always enjoy talking with her.

Anyway, we chatted over dinner – Jim was away on business – and Nancy invited me to play golf with them when he returned later that week. That's when it happened. After a good round of golf, over a couple of beers, Jim told me that he was back at Sarnoff for a brief stint to help find a new CEO and maybe spin off some of their technologies. Would I be interested in looking at some of them? "Why not?" I said, and we set a date.

Having worked at Sarnoff for eleven years, I expected a blizzard of graphs and equations in PowerPoint and an onslaught of technical jargon from each presenter. Believe me, I was not disappointed!

Just as my eyes were starting to glaze over, in walked Tom Catanese. I could sense the drive and hunger of this budding entrepreneur. He gave a succinct but spellbinding presentation on a technology Sarnoff had developed to detect "stray voltage" hazards – electrified manhole covers, street lamps, fences and the like – from a moving vehicle. Stray (or more correctly, *contact*) voltage is a major safety concern, and up to now the only way to find hazards was to check every lamp post or other metal object manually, with a handheld voltmeter.

Sarnoff was already patrolling the streets of New York looking for problems under a contract with ConEdison. Oh, by the way, the business was not only profitable but cash-flow positive and had money in the bank! Might I be interested in spinning them out of Sarnoff?

I said "Possibly...," feeling like a poker player with four aces in his hand. I agreed to go for a ride with these guys as they scanned some NYC streets at night. It didn't take long to see that they had a great business here. Before somebody else beat me to it, I quickly bought a majority share in the company, and dragged the New Jersey Technology Council venture fund along as a minority partner.

If I had been in the mood for steak that Monday night...or had politely put Nancy off instead of joining her...or had passed up on the golf date...it never would have happened. Another grace moment strikes and somehow I manage not to screw it up!

Lesson learned: Business is simple

As of October 2009 GHO Ventures had not cashed out on any of these companies. Most of them are still at least alive, if not doing well. Considering that only one out of 10 technology startups lasts for any length of time, their survival is gratifying, and there are a couple that have above-average promise.

There's been one spectacular failure, and I can lay the blame for that squarely on my own shoulders. I got enamored of a technology instead of focusing on the business and the people running it.

That was my first mistake. I don't know of a single example of a company that has thrived solely because its technology was better. But there are plenty of examples of companies that were successful because they had better people and stronger business execution.

Why did I ignore this truism? Because ideas about technology always run through my head. Most are abandoned and forgotten. But one, having to do with electricity backup, wouldn't go away. After a series of multi-hour power outages at my house in the late 1990's, I went out and had a generator installed.

Problem solved … but I remember thinking that there is probably enough energy stored in a single car battery to carry an average house for a few minutes, if not a few hours. A quick "back of the envelope" calculation revealed that 20-30 minutes was quite doable if there were no energy hogs like hot tubs and clothes dryers running. Add the appropriate electronics to convert DC to AC power, and voila! A useful system! Why hadn't somebody thought of this already?

This was the beginning of my downfall. The idea sort of drifted in and out of my thoughts over the next several years. It was probably fading away when I ran across ABC Company at a meeting of the "Jumpstarts" angel investing group. Actually, I wasn't even at the presentation. But being a member of the group, I had received it by e-mail.

When I read it, I said "Whoa! Someone has thought of it." Even though the Jumpstarts group passed on the investment, I eagerly jumped in, beguiled by the technology and without doing enough due diligence on the business and the people.

My second mistake was trying to explain away obvious signs of dysfunction that appeared shortly after GHO Ventures funded the company. It has always amazed me how people overcomplicate business. When negative behavior by the staff of a business gets in the way of progress, many investors and managers try to justify it, explain it, or work around it – instead of confronting it.

And there I was, doing the exact same thing. Most of the companies in the GHO Ventures portfolio, such as Power Survey, were busily pushing ahead with their business plans and making real progress in terms of revenues and profitability with little attention from me. By contrast, ABC's internal strife was chewing up hours and days of my time as I tried to paper over the cracks.

How bad was the atmosphere? When ABC was in its death throes, the management team was actually whining to the board that they were all underpaid, citing industry surveys to back up their claims. Underpaid? How about "underperforming?"

There were veiled threats that some of the people might leave if they didn't get big raises. Incredibly, some of the board members sympathized with these outrageous demands. But I'd finally had enough. I could only laugh at such nonsense and offer to show these people the door – especially "business development" people (sales) whose bookings had barely covered their annual salaries!

Of course there were plenty of excuses to go around: "…but these things take time…the industry isn't quite ready…you don't understand our business…we have to cut prices to meet the competition.…" If any of this sounds familiar, you've been in a dysfunctional operation (or two) yourself. And if you haven't experienced it, you're lucky.

What the board and I should have done was ask hard questions about the direction of the business – and demanded specific answers. If a company isn't meeting expectations, good managers will apologize for its poor performance, and instead of excuses, offer plans to correct the situation, including personal sacrifice and increased efforts. This was always the case at Princeton Power, which is why they did so well.

On the other hand, if the team's reaction is defensive, you're on a downward spiral that will end with a whole new management in almost every case. The question is how much time and resources you are willing to waste before you decide to act. In the case of ABC the answer was: too much.

Believe me, trust your gut and listen to it. Act sooner rather than later. When your instincts tell you that a management team isn't doing the right things – they probably aren't!

ABC was a mistake I am still recovering from. I sunk more money into it than I should have, ignored obvious signs of trouble, and when I finally examined the company instead of the technology, I had an awakening that should have come several years earlier.

Lesson learned. Big lesson learned. People either have the desire to succeed and the will to make it happen or they don't. It's as simple as that.

Business really is that simple!

Space, community, kids

GHO Ventures keeps me connected to the worlds of technology and entrepreneurship, even though I'm playing a different role. In this phase of life I'm not on the front lines, starting companies and making them work. It's a broader mission, one that fulfills the urge to keep doing something that interests me.

My life since Soyuz also includes three other areas of activity that provide fulfillment of a very different kind: staying close to space travel; supporting community organizations; and talking directly with school-children about the fun and rewards of a career in science.

Crewmate re-entry

You don't go into space and then just drop the whole concept. At least I don't. As I've said before, I'd go back up there in a minute if I had the chance. It's mankind's next frontier, and the idea of being involved in helping us explore its possibilities, even in just a small way, still has a powerful hold on my imagination.

Since it's unlikely I'll get another opportunity to blast off for the space station (to say nothing of the moon) unless I manage to create another Sensors Unlimited, I've been participating in other space-related activities instead: attending launches, going to events, conducting tours of Star City. My first post-flight space-related activity was welcoming Bill McArthur and Commander Valeri Tokarev back to earth.

After I returned from space, I sort of felt that I had left my crew-mates behind, so I vowed to get as close to their landing as possible. Space Adventures arranged for me to join the Russian search and rescue (SAR) team assigned to travel by helicopter into the bowels of Kazakhstan and find the Soyuz crew as soon as they landed.

Sometimes this is easy, sometimes very difficult. It all depends on how the landing goes. Some missions have had to do "ballistic entries," where you basically come down in an uncontrolled fashion and land maybe hundreds of miles from the planned site. Others – like the mission of Korean cosmonaut So-yeon Yi – had difficulties with the separation of modules before reentry, which resulted in their capsule going off course.

My crewmates and I were fortunate in that our two landings came in pretty much on target. The SAR team was onsite within about 10 minutes of the touchdown. These are a bunch of really dedicated guys: rough and ready, highly trained and experienced, totally committed to reaching that Soyuz vehicle just as fast as humanly possible.

The SAR team I was on started our journey the day before the planned landing, going by helicopter to a town near the city of Arkalyk, in the nearest inhabited area. We flew for about two hours over ground that was pretty much desert, flat as a pancake for as far as the eye could see. I doubt we were more than around 300 feet off the ground for the whole flight!

We slept in a "hotel" that had foam mattresses about 1" thick. Well…what do you expect in the middle of nowhere? It hardly mattered, because we were up again at about 2:30 a.m. for a quick breakfast before heading out to the helicopters to reach our assigned destination.

There are about five SAR teams spread out over the area to improve the chances of reaching the capsule quickly. I remember landing in our area just before dusk and getting out to watch the sky. All of a sudden sparks appeared above us – it was the Soyuz making its reentry through the atmosphere. It was like seeing a shooting star, but much closer up. I sort of shuddered at the violence of the sparks, and began to take my own re-entry more seriously than I had during the actual landing!

We got back into the helicopter and flew to the projected landing site. By the time we arrived, the capsule had already landed and another team was leading the crew to their designated seats and the medical tent.

The basic recovery consists of the SAR team opening the hatch and pulling the crew out of the capsule. This is because 1) the capsule is still hot from re-entry and crew members risk getting burned if they touch the sides; and 2) space travelers are often weak from their time in a weightless environment, and may have difficulty pulling themselves out.

The crew goes from the capsule right to a "lounge chair," placed close by to minimize the risk of falling down from the dizziness that accompanies re-entry.

Since Bill and I share a taste for good coffee, I had brought along a half-pound bag of freshly ground Starbucks beans, which I figured was the right thing for him on his return to earth. I sort of snuck up to him with the open bag of coffee and gave him a good whiff. He smiled and thanked me for coming to the landing. Valeri was similarly glad to see me – just imagine how I felt!

I'd also brought along two small flasks of Pinotage from my winery, which Bill and I had shared in our training days at Star City. When we got on the plane back to Moscow, I surreptitiously slipped one to each of them. But I don't think the Russians would have objected even if they saw it.

Both Bill and Valeri remember their landing as being much more abrupt than I do mine. Maybe they'd run into more wind, which would push them sideways.

Booster for Space and Science

The excitement of going into space, and the need for us to take better care of Earth while we push outward from our home planet, were both brought home during this trip. I'd started by watching the launch of TMA-8 (Jeff Williams, Misha Turin and Roberto Vitelli). Knowing what goes into this feat makes it all the more impressive.

That was a week before Bill and Valeri were set to return. I spent the time in Ukraine where, thanks to a last-minute arrangement by Natalia Shulgin, I was able to visit Chernobyl. Almost 20 years after the notorious nuclear accident there, I got a first-hand look at the place.

I won't dwell on the details of my visit, except to say that it was incredibly moving to stand in someone's apartment and actually touch the possessions they were forced to leave behind when they were told to evacuate immediately. Especially since we now know that, unfortunately, several days had passed since the accident, and they were not told in time. Anyone who doesn't appreciate the dangers we face in trying to harness the power of the atom should look at the story on Chernobyl in the April 2006 issue of *National Geographic*.

If Chernobyl is a wake-up call, space travel is a new dawn. I have been to six Soyuz launches (including mine), and I always look forward to an opportunity to see another. The two I witnessed before taking my own journey inspired both anxiety and hope. The last three have called up feelings of nostalgia: reliving a fabulous experience; seeing old friends, instructors, familiar places; eating shashliks at Star City; reconnecting with people like my interpreter Tatiana.

So it didn't take me long to accept an invitation to be a "tour guide" to a bunch of Princeton/MIT alumni who were planning to attend a Soyuz launch in October 2008. Although both universities can claim astronaut alumni, somehow I was asked to join them and show them around.

Toward the Future

Like anyone who's been successful and wants to share the good fortune, I support several charitable organizations. Believe me, I'm not looking for a pat on the back or a thank-you. I'm so glad that I was lucky enough to be able to do it.

Another one of those religious axioms goes, "to whom much is given, of him much is required." Note that it doesn't say "expected." It says *"required."* That concept rings true to me. At the end of every year I sit down with members of the charities to see where I am in helping them.

One cause close to my heart is helping children develop to their full potential. Kids are our future. I feel strongly that we have an obligation to give young people the tools they need to succeed. But while contributing money is a good thing, it's even more rewarding to get directly involved.

For 17 years I've had the privilege of serving as a mentor (that concept again) to a young man named Kevin Reeves through our local Big Brothers Big Sisters program. This group matches up kids who need an adult in their lives with carefully screened volunteers who can help them realize their full potential.

Going into space has given me the standing to expand this kind of activity. Since my launch to the International Space Station aboard a Russian Soyuz rocket in October of 2005, I've given almost 400 talks on space to school and community groups. I use my "15 minutes of space fame" to try and kindle the same kind of passion for science and technology in today's kids that my generation had after Sputnik was launched and the moon race was on.

The kids I speak to are mostly in grades 3-8. I feel that this group is the most approachable. I always wear my flight suit and play the role of a spaceman. A spaceman is what kids are looking for when I stroll into the room, and I try not to disappoint them.

I usually start by asking how many students struggle – just a little bit – with math. A couple of hands go up, and eventually most of the hands are raised. They are startled to see mine raised as well.

I tell them how I struggled with math and failed trigonometry in high school. Even after 47 years, I tell them, that stings like getting dumped by your first girl friend. Then I ask them: "So how can a guy with that kind of background end up with a bunch of advanced technical degrees, earn 12 patents, have a career as a research scientist, start two successful high-tech companies, and then go into space?"

The answer is: *Don't give up! Don't ever give up!* I tell them that those words are the secret to life, whether you want to become a scientist, a baseball player, an actress, whatever.

I relate my educational history: a 78 average in high school, conditional admission to a lesser-known college, a good but not spectacular college career. (In retrospect, I now realize that I had a big advantage over many students. I had no pretensions about being the smartest guy in the room. All I wanted was to pass!)

I tell these anecdotes not to gain sympathy or approval, but to dispel the stereotype that scientists and engineers were all raving geniuses in school, getting straight A's, never struggling (and that they were mostly white males, as well). The truth is, most struggled just like me.

Once I was giving my space talk at a middle school in the Bedford-Stuyvesant section of Brooklyn. Near the end of my presentation, one of the students shouted out "Hey…I read on the Internet that your trip cost $20 million. Where did you get that kind of money??" I was a bit taken back and embarrassed, but felt that such a forthright question deserved a clear answer.

So I told him that I had started a high-tech company that I later sold for a big pile of money, which I used to finance my trip. Naturally he retorted: "…so how'd you get this high-tech company?" I told him that I was able to start a high-tech company because even though I had failed trigonometry in high school, I didn't give up.

I'm hardly the only one spreading this gospel. The space program has provided great motivation for kids to pursue science and engineering careers. NASA astronauts are required to visit schools, and they do so willingly. I would urge every scientist to do the same. Beyond being supremely important, talking to kids in school can be a lot of fun.

There's no getting around the information, numbers, data, and, yes, equations. We shouldn't hide that – but we can make it fun for them too, and it's the fun that helps you get your points across.

To boost the entertainment quotient, I bring in some liquid nitrogen, have the noisiest kid blow up a balloon, then dunk it in the LN2…and watch their amazement as it shrinks to nothing and then fully expands when pulled out. I ask them why this happened…what happened to those air "atoms"– what are atoms, anyway…? (They're often not quite ready for the concept of "molecules.")

Then I pull out a Van de Graff generator and any kids who aren't already having a good time are pulling on my arm, begging me to let them get "charged." This is great for girls with long hair – they just delight in seeing their hair sticking straight up off their heads. During one of these sessions I usually fill up a whole memory card on my digital camera!

Many of us grew up with Sputnik, Yuri Gagarin, John Glenn and the moon race. Those were our motivators. Kids today don't have the momentous events, larger-than-life role models, and great national missions to focus their attention on science and engineering, so it is even more important now for all of us in these fields to reach out to them.

You don't have to be an astronaut to impress kids and arouse their interest. You can get that same "ohhh and ahhh" effect with magnets, lasers and prisms.

Fortunately, our government is now talking about promoting education and the sciences as a way to conquer the big problems our world faces today, just as we were inspired by Kennedy to reach for the moon. We need to get all citizens – and especially the kids – involved in science to achieve energy independence, understand our environment, and make a committed effort to put a *woman* on Mars by the year 2020.

Conclusion

Aftershocks

Looking back, my journey seems so improbable. A guy who went to college almost by accident becomes a researcher, starts and sells two companies, and winds up participating in one of the most important scientific achievements – and greatest human adventures – of all time: the beginnings of space exploration.

As a space traveler I'm one of the lucky few. But I hope and trust that that I won't be one of the last. In the future millions of people will share my experience, as the human race launches off its home planet and travels to other worlds.

That's a thrilling prospect, no question about it. And it was a privilege to be in the vanguard. But what does Greg Olsen's journey toward space have to do with anything back here on Earth?

I'm not conceited enough to think I'm some grand model for others to emulate. To me, my success is simply proof that anybody can reach for the stars. Of course each one of us has to define what that means for ourselves. For some it might be success in business, for others a career helping others, while for a small number it could be a ride on a rocket.

Whatever our goals, we can achieve them, assuming we work hard enough – and catch a break or two along the way. Regardless of our abilities, we can do more than we know.

That's certainly how it was for me. I was never named "most likely to succeed." There were always people who were smarter, more experienced, and better positioned to seize a given opportunity than Greg Olsen.

Yet I kept on plugging away, trying to get to the next level in education, research, sensor manufacturing, business, certification for space flight. Looking back, some might see this as consciously working for a better future – for myself, for my kids, for my employees, for an industry. But at the time I didn't have such grandiose ideas. I was just focused on trying to solve the next problem, build a better gadget, or make a company profitable.

It was hard work that made it all possible…and of course instances of what I call "grace," or strokes of undeserved good fortune. That's what helped me achieve success beyond anything I had ever imagined.

Maybe that's the secret to personal success. You work hard to achieve your immediate goals, and you take advantage of good fortune when you get some. You find good mentors to guide you and great colleagues to work with you. And you stay open to changing your direction when the situation calls for it.

And of course, you need the right attitude: a don't-say-die outlook on life that keeps you going when things get tough. Or as I keep saying, never give up!

One thing you don't need – or at least I didn't – is a lot of profound, far-reaching thought on your ultimate destiny. I had no grand strategic vision of where I wanted to be in two years, let alone 10. I figured out what I wanted to do in the near term, then focused on the next step in getting there.

This approach served me well in building a career, two companies, and a space endeavor. Hopefully, the same outlook will pervade the companies in my angel fund. Whether it's right for everyone, I don't know.

Finally there's the big question: what's it all for? Why work that hard to reach personal goals?

I hope my story has made it clear that reaching your goals can reward you with tremendous personal gratification, not to mention material success. But that's not the ultimate reason we keep striving for the next level. We also have an innate urge to leave a legacy for the future.

Essentially we're all preparing the way for those who follow. We have this impulse to contribute to progress. Our generation is building on what the previous generation achieved, and we in turn are broadening the horizons for our kids, so that they can see where to go next.

This sounds pretty philosophical, but to me it's common sense. If you are committed to having a positive impact in any area of human activity, you're committed to progress. Call it doing your part, call it giving back, it all comes down to building for the future.

That's what we humans have been doing since we discovered how to use fire. It's our nature. I'm glad I had the chance to participate in the most spectacular example of that impulse, the space program. I can't wait to see what comes next.

Made in the USA
Lexington, KY
21 December 2009